S0-CDK-987

THIRD WORLD CITIES

OTHER RECENT VOLUMES IN THE SAGE FOCUS EDITIONS

8. **Controversy (Third Edition)**
   Dorothy Nelkin
41. **Black Families (Second Edition)**
   Harriette Pipes McAdoo
64. **Family Relationships in Later Life (Second Edition)**
   Timothy H. Brubaker
89. **Popular Music and Communication**
   James Lull
119. **The Home Care Experience**
   Jaber F. Gubrium and Andrea Sankar
120. **Black Aged**
   Zev Harel, Edward A. McKinney, and Michael Williams
121. **Mass Communication and Public Health**
   Charles Atkin and Lawrence Wallack
122. **Changes in the State**
   Edward S. Greenberg and Thomas F. Mayer
123. **Participatory Action Research**
   William Foote Whyte
124. **Experiencing Fieldwork**
   William B. Shaffir and Robert A. Stebbins
125. **Gender, Family, and Economy**
   Rae Lesser Blumberg
126. **Enterprise Zones**
   Roy E. Green
127. **Polling and Presidential Election Coverage**
   Paul J. Lavrakas and Jack K. Holley
128. **Sharing Social Science Data**
   Joan E. Sieber
129. **Family Preservation Services**
   Kathleen Wells and David E. Biegel
130. **Media Coverage of Terrorism**
   A. Odasuo Alali and Kenoye Kelvin Eke
131. **Saving Children at Risk**
   Travis Thompson and Susan C. Hupp
132. **Text in Context**
   Graham Watson and Robert M. Seiler
133. **Social Research on Children and Adolescents**
   Barbara Stanley and Joan E. Sieber
134. **The Politics of Life in Schools**
   Joseph Blase
135. **Applied Impression Management**
   Robert A. Giacalone and Paul Rosenfeld
136. **The Sense of Justice**
   Roger D. Masters and Margaret Gruter
137. **Families and Retirement**
   Maximiliane Szinovacz, David J. Ekerdt, and Barbara H. Vinick
138. **Gender, Families, and Elder Care**
   Jeffrey W. Dwyer and Raymond T. Coward
139. **Investigating Subjectivity**
   Carolyn Ellis and Michael G. Flaherty
140. **Preventing Adolescent Pregnancy**
   Brent C. Miller, Josefina J. Card, Roberta L. Paikoff, and James L. Peterson.
141. **Hidden Conflict in Organizations**
   Deborah M. Kolb and Jean M. Bartunek
142. **Hispanics in the Workplace**
   Stephen B. Knouse, Paul Rosenfeld, and Amy L. Culbertson
143. **Psychotherapy Process Research**
   Shaké G. Toukmanian and David L. Rennie
144. **Educating Homeless Children and Adolescents**
   James H. Stronge
145. **Family Care of the Elderly**
   Jordan I. Kosberg
146. **Growth Management**
   Jay M. Stein
147. **Substance Abuse and Gang Violence**
   Richard E. Cervantes
148. **Third World Cities**
   John D. Kasarda and Allan M. Parnell
149. **Independent Consulting for Evaluators**
   Alan Vaux, Margaret S. Stockdale, and Michael J. Schwerin
150. **Advancing Family Preservation Practice**
   E. Susan Morton and R. Kevin Grigsby
151. **A Future for Religion?**
   William H. Swatos, Jr.
152. **Researching Sensitive Topics**
   Claire M. Renzetti and Raymond M. Lee
153. **Women as National Leaders**
   Michael A. Genovese
154. **Testing Structural Equation Models**
   Kenneth A. Bollen and J. Scott Long
155. **Nonresidential Parenting**
   Charlene E. Depner and James H. Bray

THIRD WORLD CITIES

Problems, Policies, and Prospects

John D. Kasarda
Allan M. Parnell
editors

SAGE PUBLICATIONS
International Educational and Professional Publisher
Newbury Park London New Delhi

For information address:

SAGE Publications, Inc.
2455 Teller Road
Newbury Park, California 91320

SAGE Publications Ltd.
6 Bonhill Street
London EC2A 4PU
United Kingdom

SAGE Publications India Pvt. Ltd.
M-32 Market
Greater Kailash I
New Delhi 110 048 India

Printed in the United States of America

Library of Congress Cataloging-in-Publication Data

Third world cities: Problems, policies, and prospects / [edited by]
John D. Kasarda, Allan M. Parnell.
p. cm. –(Sage focus editions ; 148)
Includes bibliographical references and index.
ISBN 0-8039-4484-5. —ISBN 0-8039-4485-3 (pbk.)
1. Urbanization—Developing countries—Congresses. 2. Cities and towns—Developing countries—Congresses. I. Kasarda, John D. II. Parnell, Allan M.
HT384.D44T45 1993
307.76'091724—dc20 92-30440
CIP

93 94 95 96 10 9 8 7 6 5 4 3 2 1

Sage Production Editor: Judith L. Hunter

Contents

Preface

The causes and consequences of rapid urbanization in developing countries are central issues to policymakers and academic researchers worldwide. Demographic, economic, spatial, social, and environmental change is occurring so quickly that up-to-date evidence is elusive, especially as to the net effects of rapid urban growth and the unprecedented scale of many Third World cities on individual lives and national development. With support from the U.S. Agency for International Development (AID), the Committee on Population of the National Research Council held a workshop in Washington, DC, in March 1990 based on commissioned papers addressing a wide range of significant issues concerning the causes and consequences of urban growth in the developing world. The fundamental charge given to the authors was to prepare state-of-the-art assessments of the evidence on their assigned substantive topic, emphasizing policy and research implications. During the following 12 months, based on feedback they received at the workshop, the authors revised their papers for this volume. Although the stated purpose of the workshop did not specify that the papers were to focus on mega-cities of the developing world, most authors did.

This book thus complements, updates, and extends the two volumes of *The Metropolis Era,* edited by Mattei Dogan and John D. Kasarda and published by Sage in 1988. The chapters in *The Metropolis Era* examined urban dynamics in all world regions and provided case studies of specific mega-cities. The chapters in this volume advance these discussions, giving substantially more attention to implications for development policy and needed research.

We wish to acknowledge the Bureau of Program and Policy Coordination and the Office of Population Policy of the Agency for International Development. Tom Morris from Program and Policy Coordination and Richard M. Cornelius from the Office of Population Policy played important roles in framing the issues on which AID wanted fresh perspectives. The Committee on Population of the National Research Council organized and sponsored the original workshop and transmitted the papers to AID. We thank Albert I. Hermalin and Peter J. Donaldson, respectively chairman and director of the committee when the workshop was initiated, and Samuel H. Preston and Linda G. Martin, respectively chairman and director of the committee when the workshop was held. Diane L. Goldman and Susan M. Rogers, members of the Committee on Population staff, contributed significantly to the workshop. John D. Kasarda chaired the workshop, and Allan M. Parnell was a Research Associate with the Committee on Population.

We also thank Lynn M. Igoe, editor at the Carolina Population Center, University of North Carolina at Chapel Hill (UNC-CH), for her superb editing—a word that does not adequately describe her contributions. She also prepared several of the figures and worked with bibliographic verification. Jean Elia, Administrative Director, and Patricia Zigas, International Information Specialist, both of the Kenan Institute of Private Enterprise at UNC-CH, have played invaluable administrative and research roles, giving continuity to the editorial process.

Our thanks to Mitch Allen, Judy Hunter, and Carrie Mullen at Sage for their roles in shepherding this book toward production. We are especially grateful to Stephanie T. Hoppe for her expert copy editing of the manuscript.

Francisco Alba, professor and researcher at the Center for Demographic and Urban Studies, El Colegio de México, and a member of the Committee on Population when the Workshop on Urbanization, Migration, and Economic Development was held, played a key intellectual and organizational role in the workshop.

Special thanks go to Sidney Goldstein, an eminent demographer at Brown University and former member of the Committee on Population, for his unflagging attention to the importance of demographic factors in our attempts to understand the processes of urbanization and economic development.

Introduction

Third World Urban Development Issues

JOHN D. KASARDA
ALLAN M. PARNELL

The developing world is quickly becoming an urban world. Neither civil strife, natural disasters, nor public policies redirecting industry and migrants have been able to stem this trend. Growth is occurring in urban places of all sizes from small market towns to mega-cities with more than 8 million residents.

Contemporary and projected aggregate increments of urban population in developing regions are nothing short of breathtaking. In 1950, only 285 million people, or 16 percent of the developing world's population, resided in urban places. By 1990 this number had multiplied fivefold to 1.5 billion urban residents, making up 37 percent of the total population in developing countries. The United Nations (UNDIESA 1991) projects that during the next 35 years the urban population of developing countries will triple again, reaching 4.4 billion in 2025. At that time, four of every five urban dwellers in the world will be in countries currently classified as developing, and within these countries, about two in three people (61 percent) will be urban. To place this absolute growth in perspective, simple arithmetic reveals that the *net addition* of urban population in developing countries between 1980 and 2000 will be greater than the *total* urban population in developed countries in the year 2000. During the 1990s alone, cities in the developing world will grow by an aggregate of over 160,000 persons per day (Kasarda and Crenshaw 1991).

Although rapid growth is occurring in urban places of all sizes, those of 1 million or more population have captured significant attention, if for no other reason than the unprecedented scale being reached by some of the largest cities. Mexico City, for example, had just 3.1 million inhabitants in 1950 but is expected to exceed 25 million by the end of this decade. Similarly, São Paulo, a city of just 2.8 million in 1950, is projected to reach approximately 22 million by the year 2000 (UNDIESA 1991). To view this growth in a historical context, consider that it took New York City (the world's largest metropolis in 1950) nearly a century and a half to expand by 8 million residents. Mexico City and São Paulo will match this growth in fewer than 15 years. Asia's mega-cities are likewise expanding in number and size, with the number containing 8 million or more inhabitants increasing from 3 in 1970 to 17 by the end of the 1990s.

Although smaller in magnitude, the dramatic growth of Mexico City, São Paulo, and the Asian mega-cities is being echoed in developing nations around the globe. This growth has resulted in a swift multiplication of cities containing 1 million or more residents, from 31 in 1950 to 180 today, with a projected 300+ in the year 2000. At the end of the 1990s, more than one in three urbanites in the developing world will reside in a city of 1 million or greater population (UNCHS 1988).

Accompanying the explosive growth of large cities has been a plethora of problems of seemingly unmanageable proportions. These include, among others, high rates of unemployment and underemployment as urban labor markets are unable to absorb the expanding numbers of job seekers, soaring urban poverty, insufficient shelter, inadequate sanitation, inadequate or contaminated water supplies, serious air pollution and other forms of environmental degradation, congested streets, overloaded public transportation systems, and municipal budget crises (Dogan and Kasarda 1988a, 1988b).

As a result of these worsening problems, much attention has been given to developing policies to limit large city growth by restricting migration or redirecting population and industry to smaller places. With a few exceptions these policies have been ineffective (Oberai 1987). Indeed, a number of such policies actually appear to have been fiscally wasteful and economically counterproductive (Richardson 1989b; Rondinelli 1991; Van Huyck 1988). Failure of most population redistribution policies has led to far greater recognition and appreciation by governments and donor agencies of the irreversible power of market forces, with a "swim with the tide" perspective beginning to gain dominance over "reversing the tide" (USAID 1991; World Bank 1990). Focus is therefore shifting to managing rapid urban growth rather than stemming it.

Further stimulating the market forces perspective is a heightened awareness among scholars and policymakers of the substantial benefits of large cities for individual opportunity and national economic development (Van Huyck 1988). Numerous surveys and secondary data analyses document that large cities offer migrants the greatest options for social mobility (Sachs 1988; Teune 1988). For economic development, mega-cities and other large urban places serve as highly efficient centers of specialized and essential economic activities and as primary contributors to the national economy (Richardson 1989a; USAID 1991). Even superficial analysis reveals that many of the developing world's largest cities are important global hubs of finance, manufacturing, trade, and administration. They contribute disproportionately to national economic growth and social transformation by providing economies of scale and proximity that allow industry and commerce to flourish and create needed jobs, by supporting modern education and health and social services, and by offering a wide variety of commercial and personal services essential to meeting human needs. Cities offer locations for services and facilities that require high population thresholds and large markets to operate efficiently. These cities are typically centers of innovation and diffusion and facilitate pervasive modernization (Rondinelli 1983).

In most developing countries, modern productive activities are concentrated in large urban centers, often at a level much higher than their share of national population. Thus, Abidjan, capital of the Ivory Coast, with 15 percent of the national population, accounts for more than 70 percent of all economic and commercial transactions in the country. Metropolitan Bangkok accounts for 86 percent of gross national product (GNP) in banking, insurance, and real estate, and 74 percent of manufacturing, but it has only 10 percent of Thailand's population. Lagos, with 5 percent of Nigeria's population, accounts for 57 percent of total value added in manufacturing and has 40 percent of the nation's highly skilled labor. São Paulo, with about 10 percent of Brazil's population, contributes over 40 percent of industrial value added and a quarter of net national product (UNCHS 1985).

In Pakistan gross urban product has grown twice as fast as gross rural product, and per capita urban GNP is double that of rural areas (Asian

Development Bank 1986). Nearly half of Bangladesh's total formal-sector manufacturing employment is in Dhaka, which accounts for nearly all of the country's employment in rubber products, 97 percent of jobs in the furniture industry, 84 percent of jobs in footwear production, 82 percent of jobs in leather goods, and more than half of jobs in machinery production and textiles (UNDIESA 1987b).

Lagos is the jazz capital of Africa. India's film industry, the largest in the world, is centered in Calcutta and Bombay. São Paulo and Rio de Janeiro have been internationally recognized music and dance centers for decades. As the economies of the developing world grow, the international cultural stature of their cities will likely grow as well. As late as 1970, even Tokyo was regarded as overpopulated and culturally deficient, but as its commercial importance increased, so did its level of recognition in global economic and cultural circles. Now it is acclaimed not only as a global leader in commerce but also as a world-class center of style and artistic production. Many Third World cities, some with centuries-old heritages to draw upon, are poised to follow a similar trajectory.

Often overlooked as well are the important contributions Third World mega-cities perform in transforming, distributing, and consuming large amounts of agricultural products and goods produced in rural areas, thus providing much-needed markets for the rural sector (Rondinelli 1987). More that 9,000 tons of vegetables are brought to Shanghai daily from surrounding rural production areas in China for sale in the city's markets; fruit supplies to the city have increased from about 148,000 tons in 1980 to more than 360,000 tons in 1986 (Mackenzie 1987). Likewise, more than 8,000 tons of fruits and vegetables are sold daily in Mexico City's principal wholesale market (Meléndez 1987). Given market incentives, it should be no surprise that agricultural productivity is typically highest near large cities in the developing world and drops monotonically with distance from them.

In sum, the rapidly growing mega-cities of developing countries are among the more important economic, social, and cultural centers of the world. Their importance, prosperity, and value to individual opportunity will likely continue to increase in the years ahead. These economic and social benefits, compared to what rural areas and smaller places offer, are evidenced by the hundreds of millions of residents and migrants throughout the developing world who are choosing mega-cities for home and work, despite numerous disamenities and other problems wrought by rapid urbanization.

The authors of the chapters that follow address many of these disamenities and problems as well as the opportunities and advantages Third World cities offer. In addition to assessing substantive issues ranging from migration to urban health risks, the authors evaluate issue-oriented implications for research and development policy, including conceptual frameworks that have influenced past policy decisions.

Three sets of issues are examined in this book. The initial chapters assess the theoretical frameworks that urban and migration policies are based on in light of the evidence and experiences of the past two decades. These critical assessments advance policy by pressing for the incorporation of the growing body of contemporary evidence on the processes of urban growth and the economic and social consequences of rapid urbanization. Chapters 4 through 8 cover specific substantive topics. The authors of these chapters press for the development of fresh conceptual approaches to address the demands facing institutions and individuals in the largest cities of the developing world. The final four chapters present evidence on the demographic, social, and economic complexity of processes of urban growth. Changing transportation and communication technologies have increased the movement of capital, goods, information, and people—significantly accelerating processes through which Third World cities and nations develop and change.

Assessments of Conceptual Issues

Sally E. Findley's assessment focuses on the development models that address population redistribution and urban growth. She finds the models wanting in crucial ways and proposes to adapt them by incorporating recent empirical research, particularly on migration. Findley also proposes ways to reformulate development policies to reflect the complexity of interactions between urbanization and migration. This chapter establishes the necessity of incorporating empirical research in routine evaluation of theoretical models on which policies are based.

Rapid urbanization in developing countries, especially growth of the mega-cities, has placed certain strains on urban and national economies. Harry W. Richardson presses his analysis of the role of mega-cities in development beyond the magnitude of their economic and demographic concentration, asking how much mega-cities affect national economic efficiency and the welfare of nations and their citizens. These issues are crucial to our understanding of development processes and the establishment of rational plans to address problems. Yet, Richardson finds that any analysis is constrained by conceptual, measurement, and data collection deficiencies. For example, he uncovers only limited evidence on social and individual welfare in mega-cities. Where data are available, there is evidence of both benefits, such as improved health care, and costs, specifically the negative externalities of population and industrial concentration. Mega-cities are but one result of the dynamic processes of spatial redistribution of population during development. Richardson shows that the dynamics of redistribution continue, with decentralization of urban population occurring in a number of countries.

A. S. Oberai then reviews evidence on the relationships among trends and patterns of urbanization, industrialization, and development to national economic efficiency. In his review, Oberai assesses macropolicies developed to influence spatial patterns of development. Policies implemented to promote greater rural-urban equity within a country, such as rural industrial development, often run counter to other policies promoting national economic growth. Oberai concludes that the concentration of economic activity and resulting urbanization are inevitable aspects of the development process and that macropolicies have at best a limited influence on these spatial manifestations of development.

Assessments of Substantive Issues

Inadequate formal-sector housing is one of the most visible and persistent shortcomings in Third World cities. Ellen M. Brennan examines land and housing issues in Chapter 4, showing that they are much more complicated than simply limited land and construction capabilities. Land speculation in the central city and on rapidly expanding city peripheries deters orderly development of infrastructure and inflates costs. Subsidized, high-cost housing constructed in many cities is beyond the reach of those most in need. The recent case of slum dwellers symbolically occupying a long-vacant luxury condominium building in Rio de Janeiro brought world attention to such poorly conceived land use and construction practices.

One of the most significant issues facing mega-cities is the creation of enough jobs to meet the demographically driven demand for jobs. In Chapter 5, Dennis A. Rondinelli and John D. Kasarda address the job creation needs in Third World cities using a conceptual framework that incorporates demographic characteristics, the processes of spatial redistribution of populations, and the structural changes in developing economies. Based on demographic trends in age structure alone, an estimated 228 million jobs must be created in Third World cities before the end of the 1990s. Most Third World economies have not been able to expand at the pace needed to meet labor force growth. In addition, Rondinelli and Kasarda stress that the structures of economies in developing countries are rapidly changing, requiring new skills for new economic roles, especially in mega-cities. They also discuss a range of other policy initiatives needed to stimulate greater urban job creation.

Large public and private enterprises have been unable to provide the number of jobs needed in Third World cities. In Chapter 6, Ray Bromley considers policies and proposals to benefit the development and survival of small enterprises, defined as independent businesses with fewer than 20 people involved. Bromley offers a critical assessment of current policies to promote small enterprises, focusing in particular on the International Labour Organisation informal-sector promotion strategy and Hernando de Soto's "counterstrategy," which emphasizes deregulation and privatization. Bromley stresses that any analysis or policy concerning small enterprises must consider the great heterogeneity of small enterprises in developing world mega-cities.

Exploding urban populations, strains on inadequate and deteriorating physical facilities, and social pressures to expand service coverage are all increasing the demand for public services, shelter, and infrastructure in cities in developing countries. As the same time, most Third World governments are constrained in meeting these needs by inadequate revenues and high debt levels. As a result, few national or local governments have been able to provide the services and infrastructure to meet growing needs. Overburdened ministries often provide services and infrastructure inefficiently, and many state-owned enterprises generate losses rather than revenues for the central treasury. All of these problems are leading international assistance organizations and Third World governments to reassess the potential of the private sector to take a more active role in financing and providing services and infrastructure. In Chapter 7, Rondinelli and Kasarda review and appraise privatization policies and practices in Third World cities and suggest explicit actions needed to make these policies and practices more effective.

Among the most pressing and difficult problems facing cities in the developing world is environmental degradation. Inadequate sewerage and potable water supplies and severe air pollution create serious health risks for the millions of people living in these cities. Kirk R. Smith and

Yok-Shiu F. Lee assess possible effects of air and water pollution on the health of the inhabitants of developing world cities. Smith and Lee propose an environmental risk transition framework to analyze the health consequences of rapid urban growth. This framework enables an examination of the transition from traditional risks such as infectious and parasitic diseases to modern risks associated with development. The authors focus primarily on urban environmental risks associated with air pollution and water supplies in rapidly expanding cities.

Economic and Demographic Processes Affecting Urban Growth

Development does not occur independently in each country, and external factors influence not only economic growth but also the spatial pattern of growth and thus the urbanization process. In Chapter 9, Victor Fung-Shuen Sit reviews theoretical approaches and the results of empirical studies on relationships between transnational capital flows and foreign direct investment on Third World urban growth patterns. There is ample evidence that transnational capital flows do significantly influence urbanization patterns in developing countries. But considerable disagreement remains about the processes through which these flows and foreign direct investment operate. Sit specifies a range of issues requiring additional research.

Sidney Goldstein contributes an important study of temporary migration to two Asian mega-cities, Shanghai and Bangkok. Goldstein notes that a growing body of evidence indicates that temporary migration, largely circular, "plays a key role in individual and household adjustment strategies to changing conditions at rural origin and urban destination and in the community and in national development processes generally." Circular migration meets labor demands without the "social dislocation" associated with permanent moves. As Goldstein notes, both the Chinese and Thai governments officially express concern over their population distributions, but their political and economic approaches to limiting or managing urban growth are very different. In both nations, temporary migration complements and substitutes for permanent migration.

Francisco Alba (Chapter 11) raises the important question of the role of urban growth in patterns of international labor migration and its impact on the cities of origin. The common assumption is that international labor migrants are either rural peasants or professionals in the brain drain. Alba challenges these assumptions, noting that skilled workers are increasingly involved in international labor movements, many coming from major cities. Alba shows that Third World cities, especially the giant cities, must become part of models of international labor movement as sources of migrants and as possible substitute internal destinations.

S. Philip Morgan's chapter concludes the volume by examining the interrelationships between families and urbanization, describing the increasing complexity of their interactions. Although much of the recent research on families in Third World cities focuses specifically on "urban" behaviors of families, Morgan focuses directly on the families. He examines the evidence of uniquely urban patterns of family formations, patterns of living arrangements, intergenerational support, and the division of family labor. This perspective, rooted firmly in the sociology of the family rather than in urban or migration studies, offers rich insights into the multifaceted relationships between "one of the most powerful contemporary social processes, urbanization . . . and one of society's fundamental institutions, the family."

The substantive topics covered in the chapters of this book by no means exhaust the full range of critical issues facing people, institutions, and policymakers in Third World cities. Nonetheless, the authors' thoughtful assessment of the conceptual frameworks that have spawned empirical research and development policy establishes a clear standard for extending knowledge to inform future urban research and development policy better.

1

The Third World City

Development Policy and Issues

SALLY E. FINDLEY

If population growth in Third World nations was the issue of the 1970s, population distribution and mobility are the issues of the 1990s. In the early 1970s, half the Third World countries considered growth of their metropolitan areas excessive, whereas by the 1980s almost all did so. Because of adverse developmental consequences of excessive metropolitan growth, such as unbalanced regional development, deterioration of the urban habitat, and destruction of natural resources, governments are eager to encourage population distributions that they believe will be more compatible with their development strategy (Richardson 1984c). Rural-urban migration has been considered the source of this "excessive" growth, and stopping or slowing it has been added to the set of goals of many development plans.

Despite the will to stop or slow rural-urban migration, however, rare is the government that has succeeded in "keeping them down on the farm" (Findley 1981; Fuchs and Demko 1979; Simmons 1981). The following stand out as major factors militating against success (UNDIESA 1981):

1. Weak government developmental policies (e.g., much voluntarism, but little compliance enforcement)
2. Failure of governments to perceive a conflict between excessive urban growth and national development

3. Reliance on indirect measures of control with little effort to evaluate or coordinate their effects on urbanization and migration processes
4. Emphasis on noneconomic policy levers (housing and services) rather than stronger economic policy levers (jobs and prices) or on factors influencing the salience of the economic factors—information and accessibility
5. Lack of coordination between rural and urban development strategies, particularly on control of a specific migration pattern and at different levels of the administrative hierarchy

There has been a retreat from direct interventionist policies, whether aimed at stanching rural out-migration or at limiting urban in-migration. Indirect and accommodationist perspectives are in ascendance (United Nations 1984). Increasingly, governments seek to integrate the rural and urban dimensions of their development processes to send consistent signals to would-be migrants. Much gain for internal consistency of migration and urban and national development efforts can come through better coordination and integration of the varied squares in a patchwork of development programs.

Times have changed, and we must look at more fundamental issues if integration is to occur. What are some of the changes that may affect the way we integrate urbanization, migration, and development?

- Migration forms are mushrooming, and rural-urban migration no longer dominates. International migration proliferates, often with strong consequences for national development processes.
- There are more cities and they are much bigger than ever before. Do our old assumptions about overurbanization hold in the age of mega-cities?
- Many cities have more informal- than formal-sector workers. The informal sector cannot continue to be seen as unproductive or marginal to city growth and function.
- Prodemocracy movements are sweeping the world, together with calls for more economic freedom and benefits. Governments search for mechanisms to deliver services in the short term, even as they grapple with their long-term impotency. The informal sector plays a leading role in most individual strategies for economic freedom.
- The world has become interconnected by a vast array of communications and transportation networks. The frictions of distance are much reduced and there are many more mobility options.
- The global economy is becoming transformed for global citizens, who not only buy each other's goods but see one another on television. Migrants often play key roles in actualizing the sense of participation in the world,

through their movements between regions or their purchases or economic exchanges.

The world has seen a retreat from the "can-do" philosophies of the 1960s and 1970s. The 1980s ushered in a world of caution and economic crisis, where even the goal of economic growth became elusive as many countries chalked up negative or no-growth records. Faced with impossible debt levels, many countries are involved in structural adjustment programs that severely limit the amount they can spend on discretionary development projects—the ones most likely to influence potential migrants and the urban growth process.

If governments are to integrate urbanization and migration processes sensibly into their development efforts, they need to start from an accurate understanding of current realities. Because we need to understand the current dynamics of urban growth processes correctly, I begin with the latest urban growth estimates and projections, even though they could be inaccurate because of inappropriate assumptions about rural-urban migration. I conclude with a discussion of policy options that consider these migration and urban growth processes.

Recent Third World Urban Growth Patterns

According to the most recent United Nations (UN) assessment of urbanization trends (UNDIESA 1991, 118, 131), from 1970-1990 the estimated number of Third World city dwellers more than doubled, growing from 675 million to 1.5 billion at a rate of 4.1 percent per year—three times the growth of rural populations, which rose from 1.97 to 2.57 billion at a rate of 1.3 percent per year.

Between 1970 and 1990 (Table 1.1), the proportion of people living in Third World urban areas grew 11.6 percent. By 1990, an estimated one in three Third World persons lived in a city. Although still far below the roughly three of four in cities in more developed regions, the number was up sharply from the one in four Third World people living in cities in 1970.

There is wide variation in urbanization levels among Third World countries. In 1990, 27 Third World countries still had an urbanization level of 22 percent or less—less than half the global average of 42 percent. Most of these predominantly rural countries (12 of 27) were sub-Saharan African nations. In fact, in most African nations half or more of the people continued to live in rural areas. In contrast, 32 Third

Table 1.1 Percentage of Population Living in Urban Areas by World Region, 1970-2010

	Percentage Urban					*Average Annual Urban Growth Rate*		
Region	*1970*	*1980*	*1990*	*2000*	*2010*	*1965-1970*	*1985-1990*	*2005-2010*
World	**37.2**	**41.2**	**45.2**	**51.1**	**56.5**	**2.7**	**3.1**	**2.3**
First World	**66.6**	**70.2**	**72.6**	**74.9**	**77.9**	**1.8**	**0.8**	**0.7**
North America	73.8	73.9	75.2	77.3	80.2	1.6	1.0	0.9
Europe	66.7	70.3	73.4	76.7	80.1	1.5	0.7	0.5
Japan	71.2	76.2	77.0	77.7	78.4	2.2	0.5	0.2
Australia/ New Zealand	84.4	85.3	85.2	86.2	88.1	2.4	1.3	1.1
USSR	56.7	63.1	65.8	67.5	71.2	2.6	1.0	1.2
Third World	**25.5**	**27.3**	**37.1**	**45.1**	**51.8**	**3.6**	**4.5**	**2.9**
Africa	22.9	28.0	33.9	40.7	47.4	4.7	5.0	4.2
East	10.3	15.2	21.8	29.0	36.0	6.2	6.8	5.1
Central	24.8	31.6	37.8	45.6	53.4	5.5	5.1	4.6
North	36.0	39.9	44.6	51.2	57.7	3.9	4.0	3.1
South	44.1	49.6	54.9	61.3	66.8	2.9	3.6	2.8
West	19.6	25.9	32.5	39.8	47.3	5.8	5.5	4.6
Latin America	57.3	65.4	71.5	76.4	79.9	4.0	2.9	1.9
Caribbean	45.7	53.2	59.5	64.8	69.2	3.5	2.6	1.7
Central	54.0	60.6	66.0	70.6	74.7	4.6	3.2	2.2
South	60.0	68.8	75.1	80.0	83.2	3.9	2.9	1.8
Asia	23.9	26.3	34.4	42.7	49.7	3.0	4.5	2.6
East	24.7	27.4	39.4	51.4	59.2	2.2	5.0	1.7
Southeast	20.2	24.0	29.9	36.9	44.4	3.9	4.2	1.2
South	19.5	23.2	27.3	32.8	39.9	3.6	4.0	3.6

West	43.2	51.5	62.7	70.3	74.9	5.3	4.6	2.8
Oceania*	70.8	71.5	70.6	71.3	73.3	2.6	1.5	1.4
Melanesia	15.1	17.9	19.8	23.7	29.8	8.7	3.6	4.1
Micronesia	25.7	36.9	47.2	55.6	62.0	5.8	3.8	2.3
Polynesia	32.4	33.5	37.4	44.2	52.1	4.1	3.0	2.5

SOURCES: UNDIESA 1991, Tables A.1, A.5. The 1970-1980 figures are from UNDIESA 1989b, Table 1, except for the few regions with major revisions of the series in 1990.
*Includes Australia and New Zealand.

World countries had achieved an estimated urbanization level of 67 percent or more, 1.5 times the global average urbanization level. Most of the highly urbanized countries were Latin American or West Asian. In the majority of Latin American nations, at least one of two persons lived in a city.

In most Third World subregions, the urban growth accelerated between 1970-1975 and 1985-1990. By 1985-1990 the growth rate of the urban population had risen from 3.7 to 4.5 percent per year. Only in Latin America, Oceania, and western Asia were 1985-1990 urban growth rates lower than before (Table 1.1). The rate of urbanization was over 3 percent per year in the years 1985-1990 in 21 Third World countries. Half of those with exceptionally high rates of urbanization were in sub-Saharan Africa; others included China, Bangladesh, and Indonesia.

A striking feature of the urbanization process of the last two decades has been the phenomenal growth in the number of very large cities. In 1950, only 31 cities in the entire Third World had a population of 1 million or more (UNDIESA 1980, 49). By 1970 that number had doubled, and by 1990 the 1970 figure had nearly tripled to 171 (Table 1.2).

Between 1970 and 1990, the Third World population in cities of 1 million or more had more than doubled, from 186.4 to 501.0 million. The growth rate for these large cities was 5.1 percent a year, faster than the average annual rate for the urban population of the rest of the Third World. So great was Third World metropolitanization that more of the world's residents of the largest metropolitan areas are found in Third World countries than in the more developed world. Two-thirds of the estimated 1990 population in cities of 5 million or more lived in Third World regions (UNDIESA 1991, 22).

Though one might expect that the great metropolitan areas are found in more urbanized nations, such as those in Latin America, they are more likely to be found in the nations with large populations. Half (51 out of 101) are located in Asia—14 in China. The remaining metropolitan areas are split almost evenly between Africa and Latin America. The largest number (8 of 20 African metropolises) is in North Africa. In Latin America, the majority of cities with 1 million or more inhabitants are located in temperate or tropical South America.

Table 1.2 Third World Population in Cities of 1 Million or More, 1970-2000

	Population in Millions					
	1970		*1990*		*2000*	
City Size (millions)	*No.*	*(millions)*	*No.*	*(millions)*	*No.*	*(millions)*
8.0 or more	5	45.2	13	152.2	22	319.2
5.0-7.9	6	35.5	9	60.4	12	69.6
2.0-4.9	24	66.4	52	156.8	78	240.4
1.0-1.9	29	41.4	97	132.0	59	96.7*
0.5-0.9	17	11.7	NA	NA	NA	NA
0.25-0.49	14	5.0	NA	NA	NA	NA
Total	95	203.1	171	501.0	171	725.9

SOURCES: UNDIESA 1989b, Table A.10. For 1990, UNDIESA 1991.
NOTE: In 1970 the 64 cities with more than 1 million population had a total population of 186.4 million.
*Excludes cities of less than 1 million in 1990.

More Variation in Urban Growth Rates Than Expected

Remarkably, the 1973-1974 UN projections of Third World urban growth for 1990 were quite close to the actual levels, with an average difference of less than 1 percentage point. But there are many larger if offsetting projection errors. Nine of the 15 subregions had 1990 urbanization levels lower than earlier projections for that date (Table 1.3). For this group, the average difference between the actual and projected urbanization level was 5.9 percentage points, indicating that projections were too high by an average of 14.6 percent of the 1990 urbanization level. The greatest projection errors were for Melanesia, for which the projection differed from reality by 26.9 points—139 percent. If we exclude the Melanesian projection, the average projection error for this group would be reduced to 3.3 percentage points.

Smaller average errors were made with projections that were too high. For the six subregions of the developing world with underprojected urban growth, the average error was 2.6 percentage points, for an average difference of 5.9 percent between projected and actual levels. In Latin America and Asia, actual urbanization levels were higher than

Table 1.3 Comparison of Actual and Projected Percentage Urban by Major Region, 1990

Region	*Projected 1973-1974*	*Estimate 1990*	*Difference in Points*
World	**45.9**	**45.2**	**+0.7**
First World	**74.9**	**72.6**	**+2.3**
North America	77.2	75.2	+2.0
Europe	73.3	73.4	−0.1
Japan	82.9	77.0	+5.9
Australia/ New Zealand	90.5	85.2	+5.3
USSR	71.3	65.8	+5.5
Third World	**36.5**	**37.1**	**−0.6**
Africa	35.7	33.9	+1.8
East	22.7	21.8	+0.9
Central	43.7	37.8	+5.9
North	51.4	44.6	+6.8
South	51.5	54.9	−3.4
West	28.7	32.5	−3.8
Latin America	70.7	71.5	**−0.8**
Caribbean	58.7	59.5	−0.8
Central	67.0	66.0	+1.0
South	73.6	75.1	−1.5
Asia	33.0	34.4	−**1.4**
East	33.6	39.4	−5.8
Southeast	28.1	29.9	−1.8
South	27.5	27.3	+0.2
West	63.8	62.7	+1.1
Oceania*	80.4	70.6	+9.8
Melanesia	46.7	19.8	+26.9
Micronesia	47.3	47.2	+0.1
Polynesia	47.8	37.4	+10.4

SOURCES: UNDIESA 1980, 159-162; UNDIESA 1991.
*Includes Australia and New Zealand.

expected. The largest subregional differences between projections and actual urbanization levels occurred for East Asia, where the average difference was 5.8 points, and southern and western Africa, where the average differences were 3.4 to 3.8 percent.

Projected Slowing of Third World Urban Growth Rates

By the year 2000, just over half the world's population will be living in cities. Because of wide variation in the degree of urbanization in the Third World, however, that average among developing nations will not cross the half-urban threshold until the year 2010. Even in 2010, Africa and Asia still will be just shy of the 50 percent urban mark, and four subregions will be a good deal less than half urbanized: East Africa, southeastern and southern Asia, and Melanesia (Table 1.1). Only by 2025 will all regions of the world be mainly urban (UNDIESA 1991, 106-113).

The global average annual rate of growth of the urban population peaked just before 1990 at 3.1 percent. For 2005-2010 the rate of urban growth already will have declined to 2.3 percent per year (Table 1.1). According to UN projections, the Third World urban growth rate has also peaked, at 4.5 percent per year in 1985-1990. After 1990, urban growth rates are expected to decline for all major Third World regions; by 2005-2010, the average urban growth rate is expected to be 2.9 percent per year.

There is marked variation in the patterns of decline in urban growth rates from 1970-2010. In the First World, average annual urban growth rates have been declining since 1970, but in the Third World only Latin America, central and western Africa, Micronesia, and Polynesia show urban growth rate peaks as far back as 1970. In northern Africa and southeastern Asia, urban growth rates were fairly stable between 1970 and 1990. In all other subregions, urban growth rates have been rising during the last decades. All regions are expected to show falling urban growth rates in the coming decades. A clearer picture of the shift in the pace of urbanization is seen by considering the rate of urbanization—the growth rate of the percentage urban. In 1985-1990 the rate of urbanization was 2.4 percent per year, yet only 10 years later that rate is expected to drop to 1.5 percent per year, with rates above 3 percent in only 12 nations. By 2020-2025, the Third World rate of urbanization will be 1.0 percent per year, roughly the same as the First World average in the 1960s (UNDIESA 1991, 172).

Increasingly, living in a Third World city will mean living in a very large city. In 1990, 33 percent of the Third World urban population lived in cities of 1 million or more, but by 2025 about half the world's population will live in one of 639 metropolitan areas. In 2025, only a generation away, 30 percent of the Third World urban population will live in the biggest metropolises with 4 million or more inhabitants. The big-city explosion will be most remarkable for Africa, where the number of mega-cities will quadruple from 8 to 36 between 1985 and 2025 (Dogan and Kasarda 1988a, 12-14).

By the year 2000, the number of Third World metropolitan agglomerations with more than 5 million persons is expected to increase to 34 (Table 1.2). Over half of Third World metropolitan residents will live in these very large metropolitan areas. The most astounding growth will occur in mega-cities with 8 million or more persons. In only 10 years, the number of mega-cities will increase from 13 to 22, and their populations will double, from 152.2 to 319.2 million. Hefty population increases also will be seen in cities of 2-5 million persons, which will swell by almost 100 million between 1990 and 2000.

Inadequate Models for Projecting Urban Growth and Distribution

In 1950, no Third World subregions were 50 percent or more urban, but by 2010 half of the Third World population will live in cities (UNDIESA 1991, 106-115). Because of the importance of planning for this large and growing segment of the population, urban population projections will be of ever greater interest to governments and planners. Can currently available projection methodologies provide the sensitivity and detail needed?

Given the variability of recent urban growth trends, it seems doubtful that standardized mathematical models, such as the UN urban-growth differential method, will be able to provide the sensitivity needed. For example, even though the pace of urbanization is expected to decline in coming decades, assuming this decline will characterize all subregions seems unrealistic. Looking at projections for specific regions, it is clear this assumption would be mistaken for many countries. The rate of urbanization for Africa is expected to remain high—above 4 percent per year through 2010. The very highest urban growth is expected for East and West Africa, where urbanization rates are expected to exceed 5 percent per year until 2000. So, despite the wide currentness given UN population projections, we should question their suitability for specific countries.

Hardoy and Satterthwaite (1986), in their detailed critique of 1985 UN projections, stressed problems related to baseline estimates of the percentage urban. Although it is expedient to use a standard threshold of 100,000 persons to define urban places, this threshold creates problems, as many estimates for places over 100,000 persons are based on extrapolation from much earlier census data. Hardoy and Satterthwaite urged that the figures be used with great caution and with awareness of the data deficiencies and questionable assumptions on which they are based.

To drive home their points, Hardoy and Satterthwaite showed how use of more detailed information on urbanization trends for specific cities in Latin America, Asia, and Africa can lead to a different picture from the one conveyed by UN statistics. At issue is the expected growth of the largest cities in each country. Local data suggest that the largest cities may not have the fastest growth rates. In Buenos Aires, for example, the central city lost population between 1970 and 1980, even as the suburban population of the metropolitan area grew by 30 percent. Similarly, São Paulo increased by 600 percent between 1940 and 1980, but the metropolitan area outside it grew by 1600 percent in that period. In India, the cities growing the fastest in the 1970s were either secondary cities with fewer than 500,000 persons in 1970 or those near but not in the major metropolitan areas of Bombay, Calcutta, or Hyderabad. Only in Africa were the highest growth rates found for the major cities. Between 1950 and 1980, for example, Khartoum, Nairobi, Abidjan, Dar es Salaam, and Lagos all grew by 600 percent or more. Further, in many cities, there was no correlation between population and economic growth, as is usually assumed (Hardoy and Satterthwaite 1986, 39-45). Therefore, users of UN figures are urged to do some reality testing and consider more carefully where the urban population is expected to grow and with what economic basis.

The Asian projections prepared by the United Nations also received a mixed review. Cho and Bauer (1987) compared actual and projected levels of urbanization and found that the UN figures were realistic for some countries (e.g., India, Indonesia, and Japan) but off by 11 percent for several others. Although projections appeared reasonable up to the year 2000, the UN method may lead to projections far from reality for periods farther in the future. The UN ratio-trend method works to reduce urban growth differentials between countries, which is accurate for countries like South Korea that are moving from outlier to average positions, but not for countries like the Philippines that are diverging from global trends. Like Hardoy and Satterthwaite, Cho and Bauer recommend caution in using UN projections, particularly for individual cities.

For countries with divergent urban growth patterns, it will be important to develop a variant of the urban-rural growth differential method that allows for exceptions to trends. For example, instead of assuming convergence of urban and rural growth rates for countries continuing to have high rural-urban migration, the pace of convergence could be much slower. Parameters could be introduced that control for variation in urban-rural differentials in natural increase. The current methods also assume convergence, and in many countries such an assumption is far from being correct (Findley 1991).

In the coming years, specifying the distribution of the urban population among different cities will become more important. No longer can we assume that most city dwellers will be found either in national capitals or in a few selected provincial capitals. There are more cities, and city dwellers will be spread among many different metropolitan areas.

Despite the high level of interest in mega-cities, some serious questions remain that hamper our attempts to understand and project mega-city growth. The current UN projection methodologies do not allow for marked variation in growth rates for large cities. Each city's rate depends on its previous growth rate and on the overall urban growth rate (UNDIESA 1987d, 37-38). The methodology cannot reflect sudden shifts in growth rates or major divergences between a city's growth rate and the national rate. Further, the methodology assumes an implicitly competitive model for urban growth whereby growth in a larger city reduces the growth allocated to smaller cities. Clearly, more sensitive methodologies are needed to simulate expected variations in city growth rates stemming from different patterns of economic growth or migration. If different processes underlie the growth of mega-cities, how can these processes be considered in our urban population projections? Data problems notwithstanding, can projections employ an interactive or iterative procedure that better mimics actual interurban interactions? Can other projection models be used simultaneously to forecast urban population levels and distribution, using alternative assumptions about variable city growth rates?

Demographic-economic models are an option for projecting urban growth related to internal economic growth dynamics. In these models, population growth rates are endogenous, varying with the pace of economic growth. The most notable attempt to allow for alternative assumptions about links between economic and urban growth is the Kelley and Williamson (1984) model. In this model, urban and rural growth depend on changes within the urban and rural economies or modern and traditional sectors. Although the model currently emphasizes total urban population projections, it could be adapted for subsets of cities with significantly different links between economic and population growth.

This model represents a major advance over earlier versions, but it still has its weaknesses. The parameter assumptions for relating urban economic growth and population growth are fitted on a fairly small group of countries for which data were available. Smaller and no-growth countries and ones with foreign aid dependence were excluded—the very countries for which we have the most difficulty projecting economic and urban growth. Also, the urban traditional, or informal, sector of this model may not reflect the realities found in many cities, where there is evidence of capital inputs, use of skilled labor, and perhaps most important, large apprenticeship and training functions in the traditional sectors—all of which affect informal-sector growth and its relation to population growth.

To project growth rates of different size classes of cities, the Kelley-Williamson model would have to be adapted to allow for variations in the link between economic structure and urban growth in different city size groups by selecting indicators of differential urban growth that are sensitive to variation in economic growth and pose no serious data access problems.

Wider Variation in Types of Migrants

Although UN urban population projections are not based on births, deaths, and migrations, such a component projection is much more suitable for individual city projections. We are developing increasingly sophisticated systems for estimating and projecting fertility and mortality trends, but not so for migration. Methods of projecting migrants must be improved to consider different types of migrants and their distribution among alternative cities.

Generally, we assume that the pace of rural-urban migration will slow as more and more of a country becomes urbanized, as it becomes more likely that a randomly selected migrant will come from an urban, not a rural, area. Analyses of 1960-1970 censuses for selected developing countries show that the average intercensal contribution of rural-urban migration to urban growth was 36.6 percent (UNDIESA 1980, 17). The contribution was higher for metropolitan areas, where it averaged 47 percent. In 1975-1990 projections, the migrant contribution to urban growth was expected to stay below 40 percent.

But 1980 censuses provide evidence that the migrant contribution to urban growth actually rose in several Third World subregions (Table 1.4). In Africa, the estimated migrant share of urban growth was almost double the 40 percent average seen in 1960-1970 on which the projections were based. In Bangladesh, Indonesia, South Korea, and Thailand, the 1975-1990 migrant contribution to urban growth exceeded 60 percent, also well above the expected 40 percent. Even in Latin America, where urbanization levels were higher in 1970 than in other subregions, Cuba, Peru, and Puerto Rico experienced much higher net rural-urban transfers than the 40-60 percent seen as the average net rural-urban migration contribution to 1960-1970 urban growth in that region.

The subregions with underestimated rates of urbanization tend to be those where the 1975-1990 migrant contributions to urban growth were considerably higher than expected. In Senegal, for example, the actual 1975-1990 migrant share of urban growth was estimated at 75 percent, almost double the 41 percent expectation; for West Africa, of which Senegal is a part, urbanization projections were too low by 3.8 percentage points, or 15 percent.

The 1970-1980 rise in net urban in-migration challenges the assumption that rural-urban migration rates taper off as urbanization levels rise. At least in Africa, South Asia, and Southeast Asia, there is little evidence that rural-urban migration is slowing. In these regions, rural-urban migration seems to have become a more significant factor in urban growth.

A significant problem affecting projections of the migrant contribution to urban growth is the incorrect assumption that all rural-urban migrants are permanent. The census data pertaining to migration usually refer to lifetime migration and miss circulating migrants, for whom the official residence remains the place of origin. Many migrants who live and work in the city do not move there permanently. A vast segment of migrants in the city circulate between rural and urban areas in repeated movements spanning many years, some staying for as little as a few months on each city sojourn and others staying several years. (For a discussion, see Chapman 1981; Goldstein 1978 and Chapter 10, this volume; Gould 1976; Hugo 1978, 1981; Nelson 1976; Prothero and Chapman 1985.)

Temporary or circular migration seems to be most common in Southeast Asia and West Africa (Table 1.5). In most of the countries surveyed, circulators make up 33-70 percent of the total migrant population, suggesting that official statistics may miss one to two of every three urban migrants.

Table 1.4 Estimates of Net Migration Contribution to Urban Growth for Selected Developing Countries, 1975-1990

Country	*Net Urban In-Migrant Rate/Thousand (1970-1990)*	*Urban Population Thousands 1975*	*1990*	*Average Annual In-Migrants Thousands*	*Migrant Share of Urban Growth (percent)*
Africa					
Kenya	46	1,775	5,923	177	64.17
Senegal	27	1,631	2,831	60	75.16
Tanzania	79	1,602	8,971	417	84.96
Tunisia	25	2,673	4,439	91	76.87
Latin America					
Costa Rica	23	830	1,617	28	52.59
Cuba	15	5,993	7,736	104	89.50
Guatemala	12	2,231	3,861	37	34.34
Honduras	26	996	2,240	42	50.33
Panama	9	858	1,324	10	30.73
Puerto Rico	15	1,879	2,742	35	61.04
South America					
Brazil	21	66,793	115,674	1,934	59.35
Colombia	9	11,899	22,371	149	21.35
Ecuador	23	2,984	6,136	105	49.91
Peru	21	9,313	15,681	262	61.82
Uruguay	2	2,348	2,673	5	23.17
Asia					
Bangladesh	40	6,985	15,759	455	77.77
Fiji	11	212	329	3	37.97
Indonesia	27	26,259	51,975	1,056	61.61
Iran	22	15,240	31,066	505	47.84
Iraq	18	6,764	14,034	187	38.62
Philippines	18	15,136	26,432	374	49.68
South Korea	29	16,947	31,397	701	72.77
Sri Lanka	−1	2,998	3,677	−3	−7.37
Thailand	31	6,283	12,609	293	69.43

SOURCES: UNDIESA 1988c, Table 79; UNDIESA 1989b, Table A-1.

Where circulators are 50 percent or more of the total migrant population, as in much of Southeast Asia, we need a corrected estimate of the proportion urban—the de facto urban population—which includes temporary migrants who actually reside for long periods in the city. Lacking accurate census or survey data by type of migration, one can

Table 1.5 Circulators or Temporary Migrants as a Proportion of Total Migrants, Selected Countries, ca. 1980

Country	*Proportion of Total Migrants (ranked by the highest proportion reported)*
China	72% of all migrants in nation
	40% of all rural-urban migrants to 4 cities
Philippines	71% of all migrants working in construction in Manila
	23% of all migrants, nationwide
Indonesia	70% of migrants from 4 Javanese villages
	33-61% of village population circulate
	50-85% of all rural-urban migrants, selected villages
Burkina Faso	67% of in-migrants to 2 largest cities
	60% of total rural population of sample villages
	60% of all rural out-migrants, nationwide
Malaysia	60% of all migrants from national survey
	40% of all rural households have at least 1 circulator
Kenya	57-81% in 4 high-circulation districts
	33-66% of Nairobi residents circulate
	25-41% of Western and Nyanza province out-migration
	13% of migrants to 8 cities
India	52% of rickshawalas in Varanasi
	50% of industrial workers in small city, Gujarat
Sri Lanka	45% of Colombo's workers are commuters
Mali, Senegal, & Mauritania	41% of all rural migrants from 100 villages
Nigeria	39-50% of railway workers in Eastern Nigeria
Peru	33% of rural highland village
India	29% male migrants to Bombay
Thailand	27% migrants to Bangkok are repeaters
	20% male and 33% female migrants to Bangkok
Lesotho	18% of total rural population in 5 villages
	13% of total national population

SOURCES: *China:* Goldstein 1990a; Yuan 1989; *Philippines:* Goldstein 1978; Stretton 1985; *Indonesia:* Guest 1989; Hugo 1978; Mantra 1981; *Burkina Faso:* Finnegan 1980; Piché, Gregory, and Coulibaly 1980; *Malaysia:* Goldstein 1978; Radloff 1982; *Kenya:* Goldstein 1990a; Gould 1976; Rempel 1981; Stichter 1985; *India:* Dupont 1989; Goldstein 1978; Mukherji 1985; *Sri Lanka:* Goldstein 1978; *Mali, Senegal, Mauritania:* Conde and Diagne 1986; *Nigeria:* Gugler and Flanagan 1978; *Peru:* Laite 1985; *Thailand:* Goldstein 1978, 1990a; *Lesotho:* Murray 1981.

obtain this estimate by applying the relevant proportion of circulators to the net urban in-migration rate. Thus, for Indonesia, the de jure urban population would be increased by 18 per thousand (or 66 percent of the 27 per thousand net urban in-migration rate for 1970-1980). In countries with more permanent rural-urban migration like Bangladesh, the additional temporary urban population would be closer to 25 per thousand.

If urban population projections depend on assumptions about migrants' contribution to urban growth, projections will be biased if they fail to distinguish temporal variations in urban sojourn. If circulators are counted as permanent migrants, permanent urban growth projections may be too high, but if they are omitted, their influence on urban labor dynamics will be missed.

Despite the generous documentation of a wide range of migration patterns between the rural and urban sectors, our projection methods continue to neglect temporary or seasonal residents of cities. These persons participate in the informal sector, provide cheap labor for manufacturing goods for trade on the world economy, and trundle goods around the city in all manner of conveyances. The difficulty of quantifying their numbers and economic consequences no doubt slows our ability to integrate these migrants into our models, yet we must begin to ask how much we are off by not including the populations in flux. The issue is of further import for projections based on economic growth and the contributions of the informal sector, such as the Kelley-Williamson model.

Another assumption under close scrutiny is that the capital city or largest metropolitan area dominates among destination preferences of rural-urban migrants. If this assumption held true, migrants would go elsewhere only if explicit programs were introduced to redirect them. Yet studies show that rural and urban migrants go to a range of cities. In Latin America, the largest metropolitan areas continue to dominate as centers for receiving rural-urban migrants, but their dominance is decreasing (Hardoy and Satterthwaite 1986, 42). A growing body of research shows strong preferences for smaller metropolitan areas or cities perceived as easier places to live or to find jobs (DeJong, Root, and Abad 1986; Thomas and Byrnes 1974). Not all migration flows lead in steps to the capital; some head toward regional centers and smaller cities (Drakakis-Smith 1987; Fields 1979; Rondinelli 1983).

To disaggregate urban growth by type of city, we need an alternative projection methodology that allows for different regional economic structures, variations in choice of city, and permanence of in-migration. To keep the projection model within manageable bounds, we need agreement on the different categories of cities, economies, and migrants for which the models are to be used. This means focusing on specific regions or types of cities. Given the detailed knowledge of the city required, these projections can best be prepared at a local level. Where available, the UN projections can serve as a baseline projection against which the variants are compared.

In summary, if attempts are to be made to synchronize national and urban development processes better, the methodologies for estimating, projecting, and evaluating urban growth processes must be updated. First, these methodologies must be sensitive to the diversity of residential patterns in cities so that transients, circulators, and the "doubled up" populations are included. Second, we must question our theories about "proper" urban population distributions given the global importance of mega-cities, changes in interurban hierarchical interactions, and so on. Third, models for projecting urban growth must be rethought to consider emergent concerns: regional differences in urban growth rates; differential economic bases for urban growth, including urban-rural and urban-global linkages; and the importance of the urban informal sector. With improved methods, we can try to evaluate how much the hard choices about service allocation, infrastructure, and other resources may have influenced the process of urban growth.

Shifting Perspectives on Appropriate Urban Population Distributions

It seems unlikely that the "turnaround" dynamics evident in Europe, Canada, and the United States will appear in the Third World, where high growth rates in the largest cities run counter to First World experiences. (For discussion of the urban turnaround, or deconcentration, see Korcelli 1984.) Richardson (1984c) maintains that large city diseconomies—congestion, pollution, and high costs of living—will drive Third World urban deconcentration, just as in the First World. His arguments are supported by data from São Paulo, where environmental problems, congestion, and commuting problems have slowed growth (Hardoy and Satterthwaite 1986). Rondinelli (1988) suggests that the same forces will drive a modified form of deconcentration in Africa. But Vining (1986) provides counterevidence; his analyses of 1970-1980 population data for major metropolitan regions of 44 countries yielded only limited evidence of deconcentration. This debate is still unresolved, but it serves to remind us to consider carefully assumptions made about total versus disaggregated urban populations. The debate over concentration versus deconcentration is part of a larger one on the validity and utility of the concepts of overurbanization and primacy. This debate concerns the most appropriate distribution of the urban population among different-sized cities. Some have argued in favor of a rank-size rule, whereby ideal size for a city is the size of the largest city divided by its rank. For example, the second largest city should be half the size of the largest city. According to this theory, if the largest city is more than twice the size of the next largest one, then it overly dominates urban exchanges and markets. When the largest city becomes several times bigger than the next three cities, it is defined as *primate,* and urban growth in the other cities will be stunted, with negative repercussions for the relations between these cities and their rural hinterlands.

The primacy and overurbanization arguments stem from assumptions about urban economic growth. A basic premise is that marginal production costs rise more steeply than average costs in very large cities, whereas marginal production benefits fall more steeply than average ones (Mera 1981). This makes it less profitable to do business in the largest cities. The benefits of a more diversified labor force and a larger potential market for goods, namely the agglomeration economies, are unattainable because of high commuting costs and other costs of doing business in a city. As a result, the marginal costs of production in the very large city exceed marginal benefits. Thus, in the interests of balanced urban and regional development, countries are encouraged to avoid primacy or the overurbanization associated with top-heavy urban systems. (For a discussion of these arguments, see Richardson 1984c.)

Others claim the rank-size rule has no relevance today. How can we compare such dissimilar concerns as costs of pollution, congestion, and crime against the benefits of concentrated and relatively healthy urban markets, industrialization potential, and a higher quality of life for many metropolitan residents? How can any decision about "optimal" distributions be made when the dynamics of economic growth vary from city to city? What is the meaning of primacy when we live in an age of interlocking urban economies, instant global communications, transnational corporations, and world trade areas? And how relevant is the concept of *overurbanization* in a world of mega-cities? In addressing these questions, the following points have been raised.

The diseconomies of scale argument has lost much weight. Analyses do not allow for increasing returns to scale in the cost of providing services and infrastructure. Firms locating in the city may not have to pay all the costs of pollution, congestion, and so on. Average benefits also may increase with city size: diverse tertiary services, efficient transportation and communications services, enriched universities, and so on. These are seldom accounted for when evaluating the benefits of large cities. There is no evidence that the overall marginal benefit of population increase in large cities declines across a range of city sizes (Mera 1981).

There is no strong evidence that in the 1990s Third World cities derive their economic growth from industrialization. Thus, arguments based on a historical link between levels of urbanization and industrialization are no longer relevant (Berry and Kasarda 1977; Oberai 1989). Signs of heterogeneity in production processes and variety of the economic engines driving city growth would necessarily alter the arguments about primacy. Probably due to this heterogeneity, the evidence is mixed for a link between population distribution and economic growth. Using 1960s data, Mera (1973) showed a positive relation between primacy and economic growth rates, but an analysis using 1970s data failed to confirm it. Richardson (1981) ranked 36 Third World countries by per capita GNP, and despite wide variety in degree of urban primacy, urban population growth, and rural-urban migration, he found no proof of a regular link between population distribution and economic growth.

Further, it is no longer reasonable to assume that there is only one big city per country. Although most Third World countries can count on at best one mega-city, countries increasingly have two or more metropolitan agglomerations instead of one large city and several much smaller ones. The issue becomes less one of slowing growth of a particular metropolis and more one of fostering alternative growth patterns among a few giant cities and several secondary ones. A larger pool of giant and secondary cities gives more choice and thus flexibility to urban planners trying to reduce concentration in a single metropolis.

A major criticism of the rank-size perspective is its inherent isolationism, since urban growth is seen as if it were isolated in time and explicable only by other internal processes (Timberlake 1985). In reality, urbanization is shaped by processes extending far beyond national borders. According to the world systems perspective, the pace of urbanization is set by the city's role in international capital accumulation processes (Meyer 1986a; Portes and Walton 1981). Therefore, primacy or overurbanization labels are meaningless, completely dominated by the much more powerful links between core and periphery (Clark 1986; C. A. Smith 1985; D. A. Smith 1985).

If cities are the sites of international economic transactions, then they must compete with each other in the world economy. World-class cities need world-class levels of infrastructure: reliable, sufficient electricity; ample water and sewage treatment facilities; telecommunications; transport networks in and beyond the city, especially air services; financial and professional services; and a cultural and educational infrastructure for a highly trained labor force. These are very expensive, especially for small cities without economies of scale or markets to support such services. Trying to spread these among multiple cities in a country would be inefficient, if not impossible. Singh (1989) concludes that on efficiency grounds, there is little justification for decentralization, as few modern industries or services can truly be shifted into smaller cities without real economic losses. Under this perspective, each country's capital or largest metropolises should be strengthened so that they can compete better in the world economy.

The primacy/overurbanization debate was framed before the communications and transportation revolutions. These have diversified communications media and increased distances from which people are drawn into the urban labor market, much altering the calculus of diseconomies of very large urban agglomerations (Hugo 1981; Teune 1988). These technologies decrease the friction of distance associated with large metropolitan size, even as they augment problems of size by effectively increasing the city borders to include the band of villages sending commuters to city jobs. Cities are enabled to diversify their spatial structures, and in most large metropolitan areas the single central business district has given way to a multinucleated structure, so that all the city's businesses and services need not concentrate in one area.

Considering these issues, it seems more realistic to set aside the issues of primacy or overurbanization per se and instead focus on urban concentration versus deconcentration. This reflects a view that *overurbanization* is a term developed for understanding First World urbanization processes that have little relevance to Third World urbanization phenomena, where the links between economic growth and urbanization are far more varied and complex (Berry and Kasarda 1977; Drakakis-Smith 1987; Dwyer 1986; Moir 1977; Oberai 1989).

Accommodationist Versus Interventionist Perspectives on Large Cities

Policy implications of the deconcentration/concentration perspectives are seen in the choice between accommodationist or interventionist strategies. The accommodationist views concentration as not only inevitable but positive, so policies should be designed to foster the positive aspects of the concentrated urbanization. The interventionist views unbalanced urban growth as detrimental to national development. To promote balance, governments should channel population to the cities with the most promise for realizing agglomeration economies and positive trade between the cities and their rural hinterlands.

When considering long-term urban development trends, the accommodationist and interventionist perspectives converge. Proponents from both perspectives propose that systems of cities over a period of generations go first through a period of concentration, then to primacy, and finally to deconcentration. Richardson (1984c) argued the "natural" inevitability of the process, and Meyer (1986a) further substantiated it in an empirical analysis of city size growth patterns in newly industrialized countries. Where the accommodationist and interventionist perspectives diverge is on timing of the deconcentration phase and government's role in managing this process: Accommodationists would follow the trends; interventionists would lead or steer them.

The accommodationist views cities as engines of economic growth. To foster such growth, mechanisms should be developed in large cities to facilitate economic innovation and expansion. According to Teune (1988), the city uniquely promotes variety and provides access to it through its communications and transportation infrastructure. The appropriate size for a city is given not by some arbitrary figure but by the communications and transportation infrastructure, which determines the niche the city can grow to fill.

Teune further suggested that big-city pathologies (pollution, epidemics, crime, etc.) have been much overestimated and are based on very limited data that ignore the individual and family benefits of urban life. Studies of self-help housing and community-based sanitation or waste disposal systems provide examples of ways private individuals and communities internalize their service and infrastructure costs, thereby reducing the public costs associated with living in the city (Bromley and Gerry 1979; Robinson 1987).

The accommodationist is much more likely to seek ways to foster positive contributions of the informal sector. This view is supported by evidence of economic innovation and production from indigenous, nonindustrialized areas of cities in Latin America, Africa, and Asia. A detailed study of the evolution of indigenous industrialization in textiles and leather goods in Guadalajara, Mexico, led Arias and Roberts (1985) to stress the importance of a strong informal production sector to enable full use of family labor. In his overview of African cities,

O'Connor (1986) also concludes that cities serve a critical function in giving access to production opportunities not present in rural areas. The large African city becomes the meeting ground for diverse ethnic groups, where there is a chance for wealth and privilege to spread. Bromley and Gerry (1979), Deble and Hugnon (1982), Hugo (1987), Jules-Rosette (1985), and Peattie (1975) stress the individual benefits of informal-sector work that drives rural-urban migration and keeps migrants in cities, essentially voting with their feet for concentration.

Some suggest viewing the city as a theater for accumulation in which it becomes possible for large numbers of entrepreneurs to succeed. In the large cities, small and large enterprises alike can profit from the most diverse and vast support systems. These cities are centers for modern commerce, foreign and local capital markets, and diffusers of modernization, efficiency, and other norms linked with Western culture (Armstrong and McGee 1985).

The accommodationist recognizes that explicit attempts to alter the migration patterns or growth rates of specific cities have seldom had their intended effect (Laquian 1981; Simmons 1981). Also, deliberate interventions in migration flows or residential choices are seen as restricting individual freedom and disturbing the automatic equilibrating systems that naturally link the distribution of population and economic opportunities (Stöhr 1981, 43-45). Although there are severe strains on the infrastructure of some large cities, such as Lagos and Kinshasa, restricting mobility would be counterproductive, as urban and economic development go hand in hand.

Instead, supportive urban development policies are needed, such as more effective urban public administration, progressive revenue instruments, and planning for a more complete absorption of the urban labor force (Montgomery and Brown 1986). Migrants should be viewed as part of the solution, not part of the problem. For example, migrants could be put to work building needed low-cost housing and providing services (Robinson 1987).

Accommodationists figure prominently among those calling for stronger policies to incorporate marginal populations in Third World mega-cities (see Dogan and Kasarda 1988a; Perlman and Schearer 1986). Knowing that Third World mega-cities will increase substantially in the coming decades, their proponents urge better understanding of the processes of accumulation and economic growth in them so that we can foster economic growth and explore creative and cost-effective ways to accommodate population growth in these cities that are seen as vital to national economic growth (Kasarda and Crenshaw 1991).

In contrast, the interventionist considers big-city problems to be unsolvable. Cities that grow disproportionately large no longer stimulate regional economic growth because their economies become parasitic. Excessively large cities exploit the hinterland, leaving it impoverished and dependent (Hoselitz 1955). Concluding that primacy does not generate positive economic benefits, interventionists agree that overconcentration is harmful and to be averted (see, e.g., Berry 1971; Friedmann 1968; Rondinelli 1983; Ternent 1976).

There seems to be consensus that the size range at which intervention is most appropriate is 1-2 million, after which coping with urban population growth shifts to a different plane (Galantay 1986). At this size innovation potential may be greatest, but thereafter the ability to carry out innovations is increasingly constrained by breakdowns in the transportation or communications systems.

The interventionist argues for a more balanced and interacting system of cities. With such a structure, cities can stimulate economic growth in the hinterland and the nation as a whole. The goal is to create the positive features of urban economic innovation and expansion in a more dispersed and workable network of cities (Hansen 1981; Rondinelli and Ruddle 1978). Although interventionists agree that government should intervene to channel urban growth better, they disagree on the best strategy. Some emphasize rural-oriented interventions; others stress urban-oriented plans (Findley 1981; Simmons 1981). The difference lies in the researcher's analytical perspective. Some interventionists consider the main problem to be the way large cities exploit their peripheral hinterlands. The rural periphery produces food for the cities, export crops to earn foreign exchange, and labor for the urban industrial and service sector (Johnson 1970; Lipton 1982; Standing 1985). When the city takes more than it returns, it only adds to rural poverty.

Some empirical studies demonstrate a negative relation among export crop dependence, primacy, and slower national growth. Using 1980s data, Clark (1986) found a strong negative link between urban primacy and economic development in newly industrialized countries. Frey, Dietz, and Marte (1986) found a small, positive relation between dependence on exporting raw materials and primacy. London and Smith (1987) show the connection among all three; their regression analysis of data for 80 nations shows that investment biases favoring large cities and export dependency have simultaneous and negative consequences for national economic growth. Jansen and Paelinck (1981, 31-46) report slightly different findings. Using 1960-1975 urbanization rates and levels, they found no strong relation between the urbanization rate and trade dependence but a strong negative relation between primacy and dependence on primary sector trade in the world market. These findings suggest that the problem with primacy is not necessarily size per se but the dependence of metropolitan growth on single crops or primary products.

According to Portes and Walton (1981), urban elites derive part of their power through cooperation with rural elites, and together they foster neocolonial and dependent rural development. Changing rural and urban development patterns thus requires strategies that alter urban and rural elite power structures and their control of foreign capital. Interventions should focus on enhancing rural political and economic power and revising the terms of trade to provide rural people a better deal. Doing so is expected to reduce rural-urban migration and slow the inevitable urban concentration and rural impoverishment. Interventionists argue that the one way to reverse this is through alternative production strategies, especially those that develop other regions.

The changing nature of the industrialization process makes intervention more acceptable economically than it was in the past. In many countries, but especially in African countries, industrialization is not the major source of national economic growth; the rural sector is, so primate city investment is probably irrelevant to economic growth. Evidence is scant for a trickle down of primate city industrialization to other cities or regions. Even if urban economic growth does spread out from the city, evidence is mounting that the strongest growth sectors in Third World cities are small and medium-sized enterprises, which can as easily, if not more easily, be located in small or medium-sized cities. Further, marginal employment absorption capacity declines rapidly in metropolitan areas of 1 million persons or more (Rondinelli 1988). All these factors suggest that there may be just as many, and perhaps more, economic returns to investing in industries or enterprises in secondary cities.

High costs of urban growth and management in primate cities are other major justifications for seeking to channel urban growth to locations less costly to service. For example, O'Connor (1986) and Teune (1988) argue, in contrast to Montgomery and Brown (1986), that the severe urban pathologies of Lagos, Kinshasa, and other large, overgrown cities are compelling evidence of the difficulties of trying to maintain some modicum of services when the city grows far beyond original expectations and its own economic base (Lazlo 1986).

Depending on the problem diagnosed (rural exploitation versus inability to absorb growth), interventionist policies take two forms: redistributive interventions explicitly to influence population distribution to alter the rate or direction of social change and interventions that adapt population distribution to the consequences of societal change, such as the emergence of spontaneous growth centers (Stöhr 1981, 42-45).

Interventionist strategies use a mix of policies. Some stop migrants from entering or working in a certain city; some provide incentives to factories to locate in second-tier cities; some create infrastructure needed to support migrants or firms coming to newly developing cities or regions. The goal is not to create a wholly different set of communities but to channel urban growth and migration so the hierarchy of urban settlements works with equity and efficiency. The interventionist does not see an entirely new urban world but one offering more choices for people outside the biggest cities.

Dennis Rondinelli (1983) offer a concise plan for diffusing urban growth among second-tier cities. He suggests that programs to strengthen secondary cities relate to specific functions these cities might serve in diffusing economic growth: public and social services, small-scale manufacturing, commerce, regional marketing, agricultural processing and trade, off-farm employment, transportation and communications, and formal and nonformal education. This is essentially a policy of decentralized concentration, or seeking out and reinforcing the nodes where concentration can do the most good for necessary development functions.

Case studies from Brazil and Mexico illustrate the realm of changes that might be achieved through effective interventions to balance city size and economic activity better. The Brazilian case emphasizes the need to focus on production relations with the hinterland. Faissol (1986) notes that São Paulo is an example of a new primate region that has 60 percent of the total population but 90 percent of the industrial output. Brazil's older primate city, Rio de Janeiro, has increasingly assumed a distant second place. Lesson 1 is that nothing is sacred about a historical size advantage; changing linkages to the world economy can strengthen other cities. Lesson 2 is that merely fostering growth of the second city does not necessarily eliminate the unequal distribution of economic benefits assumed to be associated with primacy. If anything, the São Paulo metropolitan region portrays a more unequal distribution of wealth than Rio. In São Paulo, industrial elites have emphasized the consumption of imported luxury consumer goods, undercutting the demand for cheaper alternatives that could have been locally produced.

This reduces the multiplier effect that normally would have been expected with the high level of value added in the region. Diffusion of development is advanced in São Paulo, compared to northeast Brazil, but not as advanced as it might be under alternative consumption and diffusion processes. In contrast, the regional city of Porto Allegre, smaller but still over 1 million, has an economy that generates a much larger number of jobs across the income range. Faissol suggests that the regional spread effects of secondary city growth cannot be assumed as an automatic correlate; conscious efforts are needed to facilitate the spread effects into the wider urban and rural economies.

Guadalajara, Mexico, is a city that has succeeded in maintaining a stable, well-paid labor force—an indicator of balanced growth. Small-scale firms producing consumer goods have always dominated the Guadalajara economy, and flows between the urban and rural economies reflect the strength of this specialization. These ventures employ women and children and allow a greater labor market turnover than could exist in a more limited, strictly industrial labor market. City workers support themselves by combining industrial earnings with the security wage of women and children working in the small-scale firms. This scheme provides migrants and natives with a stable income base, which in turn permits a more stable, less dependent relationship between the city and surrounding rural region. Instead of being forced to serve as a labor market reserve and retreat for the unemployed or displaced, the rural region is a strong market partner for Guadalajara's enterprises (Arias and Roberts 1985).

But it seems that for every example of a city that has somehow managed to generate jobs and interact positively with its rural hinterland, there arises an unsuccessful case. Lack of information about cities that work makes it harder for the interventionists to overcome the accommodationist view that intervention won't work. The paucity of information on successful strategies for maintaining or stimulating healthy city growth also limits the interventionist vision. More concrete studies of alternative urban development strategies would surely be helpful.

World Systems Perspectives on Interventions in Urban Development

Some argue that the logic for intervening (or not) does not follow from generalized rules or formulas for ideal city size distributions. Rather, each nation has a unique urban dynamic, influenced by its place within the regional class structure. Changing population distribution requires far more than urban planning and infrastructure management. It calls for a transformation of the economic system, and especially of the class structure allied with control of the means of production. Proponents of this view argue that the linkage of interest is that between the city and its production process and the broader world economy. In this light, urban development strategies are inseparable from global development processes.

A leading exponent of the world systems perspective, Michael Timberlake (1985), points out that most urban development theorists view cities as if they were isolated in time and space. Despite models based on an ecological perspective, urban system boundaries are usually drawn at the edge of the city, region, or nation. But in the current interlocking, capitalist world economy this characterization is inaccurate. Urbanization does not simply reflect an internal, spatial division of labor, but global patterns of labor and resource exploitation. The urbanization process cannot be viewed in isolation from the much larger world system.

Manuel Castells (1977) argues that the city is a projection onto space of the mode of production and its social organization. Thus, each production system would be associated with a unique urban system. Most of us are familiar with urbanization as an expression of capitalistic industrialization. Urban systems differ because of the influence of earlier production systems or of those that may operate beside the capitalist system. As the world goes through cycles of accumulation, competition, and exchange, urban systems will change.

If urbanization processes reflect capitalist cycles of production and accumulation, understanding and influencing them requires an analysis of the class structure of urbanization. Imbalances in the urban system cannot be changed simply by tinkering with the composition of industries or other peripheral characteristics; social transformation in the way elites interact in the world system is required, as these interactions affect a nation's urban and rural development processes (Portes and Walton 1981).

The Issues Before Us: The Nexus of Migration, Urbanization, and Economic Development in the 1990s

In the 1990s, the issues raised here must be examined carefully so that our conceptual frameworks and methodologies can reflect the realities of migration diversity, large-scale urbanization, and complex economic linkages that span regions, countries, and the globe. Based on this review, the most critical issue seems to be the link between alternative economic development models and the patterns of migration and urbanization that became dominant in the 1980s.

The first set of research questions centers on the need to reexamine our assumptions and models for projecting urban growth. The problem is that rates of growth have been much more varied than expected. Certain cities in certain regions grew much faster than expected, and our inability to predict this variation is most unsettling. Urban population projection techniques need to be developed that do not assume convergence, particularly at a subnational level.

We need to modify estimation, projection, and urban growth models to take into account current realities of migration. Permanent rural-urban migrations no longer dominate international and internal migration patterns; in Latin America and Asia, for example, permanent rural-urban migrants may constitute less than 20 percent of the total migrant flow. To measure and evaluate properly the role of the diverse forms of migration in the urban development process, methodologies need to be refined to capture the space-time dimensions of migration better. More experimentation is needed to develop definitions using place of work, de jure versus de facto residence, relation to household head, or number of moves in the last year. Just as census-taking methodologies are being modified to obtain a more accurate count of the homeless in First World cities, we need better methods to count the floating and shifting Third World city population. Only with these more sensitive methodologies can we reevaluate our assumptions about the migrant contribution to urban growth.

Among the most challenging tasks is the need to consider differential migration rates, relative to the size and economic structure of the urban destination. No longer should we assume that all migrants end up in the largest city, yet we have no multidimensional migration-urbanization typology that can help us categorize the different patterns. Just as migration typologies need to be refined to allow for varying degrees of insertion into cities, we need to cross-classify these migrant patterns by type of city. Indeed, some researchers suggest that the patterns are simultaneously determined and that it is impossible to develop the migration categories without considering the destination characteristics and niches available to migrants.

A second set of questions concerns our assumptions about "appropriate" urban population distributions. Increasingly, living in a Third

World city means living in a very large city. In 1990, 33 percent of the Third World urban population lived in cities of 1 million or more. Despite the sometimes horrific living conditions of their poor, the largest cities, including their metropolitan fringes, continue to provide a broader range of economic opportunities than can be offered in some of the smaller and more specialized cities. We must stop equating *primate* with *bad* and adopt a more accommodationist perspective.

At the same time, there is evidence that letting big-city growth run its course unimpeded can exacerbate the already enormous challenge to national planners of servicing the needs of the urban population and fostering balanced, integrated national development. Needed here is a more pragmatic blending of the accommodationist and interventionist perspectives. We must work together to specify where policies to alter urban growth and in-migration rates would be desirable and feasible, given other possible development interventions.

There is no consensus on a recommended distribution, as there is no single "good" distribution. Much depends on a country's developmental style, its institutional capacity, its level and strategy for economic development, the roles of the informal and formal sectors in that process, the balance of urban and rural economic contributions, and so on. How can governments move beyond the rejection of massive rural-urban migration to a strategy reflecting the realities of urban growth processes and alternative migration systems? Are there urban population distributions that foster the transition from dependent, nonindustrialized to newly industrialized? What do the Brazilian, Korean, and Thai cases tell us? Which types of migration have fostered these transitions? Are these situations applicable elsewhere?

In considering alternative urban growth and population distribution patterns, we need to go beyond economic growth rates to consider a broader range of development goals, such as equity, welfare, political stability, quality of life, efficiency, and environmental protection. We need to push our data collection and estimation methodologies to provide more subnational and regional indicators of these other dimensions of development that necessarily enter into the evaluation of alternative urban growth scenarios.

To address concretely the issues of where, when, and how to intervene to complement or channel urban growth processes, we will certainly need to reconsider our models relating urban population and economic growth. The Kelley-Williamson model (1984) is a significant step in this direction, but additional work is needed to include more realistic demographic assumptions and to create parameters that adapt the model to a wider range of cities (including small and no-growth cities).

In addition, more study is needed on the links among urban population growth, national development, and global or supranational development processes or constraints. To what extent does the worldwide economic system constrain efforts to facilitate specific internal distributions of population? What are the consequences of a stagnant world economy or fluctuating fuel costs? What are the implications of the communications and computer revolutions? What about debt restructuring? How can countries balance the need to shelter and feed their urban populations with the often competing goals of minimizing further destruction of prime agricultural lands or forests?

A blending of accommodationist and interventionist perspectives also depends on a more realistic view of the informal sector. Current research shows that it is far from a residual or nonproductive appendage to the dynamic modern or formal sector. The informal sector encompasses a wide range of secondary and tertiary activities, and plays a key role in facilitating capital accumulation and in adapting to changing price, labor, and input supplies. We still do not know, however, the dynamics underlying the search by migrants and urban poor for niches in the informal sector and the balancing act between the informal sector and other sectors, within the city and beyond it. How do people choose a niche? How is this choice influenced by individual attributes or objectives, as compared to the structure of the urban economy and environment and its links to the rest of the region?

These questions are intentionally general and global, but this does not mean that the research must be equally abstract or global. On the contrary, we need focused studies of everyday life in a variety of urban settings. In conducting these studies, we need to probe continually for the micro-macro linkages between the lives of individuals and the structures of the cities in which they live. The micro-macro linkages perhaps can be best addressed in comparative studies, orchestrated to include the participation of different disciplines and targeted to focus on contrasting regions. By including among these comparisons regions or cities with contrasting development intervention strategies, we will get closer to the goal of understanding how specific interventions or policies might contribute to shaping migration and urban growth patterns.

2

Efficiency and Welfare in LDC Mega-Cities

HARRY W. RICHARDSON

Mega-cities—defined in this chapter as urban agglomerations of 8 million or more—play an important role in less developed country (LDC) urbanization even though their geographical distribution is very uneven. For instance, in many but not all cases they are primate cities with key systemic functions in national, or even international, urban hierarchies. They may account for a huge share of the national urban population, such as the 69 percent share of Bangkok in Thailand. Their role in modernization is crucial, although their economic structures may continue to reflect the paradoxical dualism of their host economies, combining modern industrial bases and postindustrial service functions with massive informal-sector contributions. They remain critically important in rural-urban transformation because they absorb, directly or indirectly, large shares of the rural surplus labor migration streams, as evidenced by attempts to divert them to secondary cities and towns that fail because these smaller urban centers lack sufficient absorptive capacity.

Yet, analysis of the role of mega-cities must look deeper than their shares in urbanization and in economic activity. The real test is how much they contribute to economic efficiency and, even more important, to societal and individual welfare in the countries where they are. Given that there will be only about 30 mega-cities worldwide in the year 2000—only 23 in developing countries—most countries in the world lack a mega-city, and many do quite nicely without one. A large national population size is obviously a necessary but insufficient condition; Peru is the smallest country with a mega-city, and its projected 2000 population is about 28 million (UNDIESA 1987d, 232). Thus, unless mega-cities are to be studied in isolation as purely urban phenomena rather than as economic entities that are merely one, if an important, component of a geographically larger set of production and consumption activities, the mega-city story is essentially one part of a much larger and broader theme—the economic, social, and political present and future of large, but not necessarily rich, economies. Even in these countries (developed and developing), mega-cities account for only 8.2 percent of projected combined national populations in the year 2000 (based on projections in UNDIESA 1987d). To place the mega-cities in an even broader perspective, they will account for about 12.7 percent of the world's urban population in 2000, or 5.9 percent of the world's total population (UNDIESA 1987d). These are not negligible proportions, and the percentages would be much higher if income, output, or other economic measures were used, but they make the point that mega-cities are only one strand in the economic development weave.

But numbers and proportions may underestimate both the positive and negative contributions of mega-cities to the economies and societies of which they form a part. Mega-cities may be critically important as innovation sources or as the initial receptors for imported innovations; this role may be even more crucial for social or political than for technological innovations. For those who believe that cities are the cradles of civilization and culture, the mega-cities may be especially important for the fusion of national and international ideas. Productivity improvements in industry, even in agriculture, may first be initiated, if not applied, in mega-cities. Physical and especially human capital are overwhelmingly concentrated there. The negative externalities associated with the spatial concentration of population and economic activity, such as traffic congestion, pollution, pressure on scarce services, and exposure to crime and social unrest, are widely assumed to be more prevalent in mega-cities than in smaller urban areas. For some observers, these negative aspects of spatial concentration are so overwhelming, so far outweighing any economic benefits flowing from agglomeration economies, that mega-cities are labeled as examples of *urban pathology, gigantism,* or *macrocephalism.*

The Growth of Mega-Cities

The United Nations (UNDIESA 1987d) is the only organization or research group that consistently prepares estimates and projections for mega-city populations (using the same projection method for all cities). The data are not without weaknesses; despite the adoption of the concept of *urban agglomeration,* some of the estimates are subject to "underbounding." Even sizeable differences may not be critical for mega-city analysis; the problems of a city with a population of 12 million are not inherently different from those with a population of 11 or even 10 million, although the differences may be very important for local planners projecting the demand for local public services.

In Table 2.1 I show UN estimates and projections for the world's 30 largest cities, developed and developing, from 1950-2000. This is probably the full list of mega-cities defined by an 8 million cutoff, although Bangalore might squeak in, and several others should grow beyond 6 million (Hong Kong, Bogota, Pusan, Lahore, Chicago, perhaps Ankara and Kinshasa). These data suggest several generalizations. First, mega-cities are a modern phenomenon. Using the 8 million threshold, there were only 3 mega-cities in 1950, although the number rose to 14 by 1975. Second, it is a developing country phenomenon; only Moscow and Milan from outside the developing world have joined the ranks since 1975. Third, in many developing country mega-cities, the pace of growth since 1950 is astounding: a 26-fold increase in Dhaka, 23-fold in Lagos, 19-fold in Baghdad, 12-fold in Seoul, over 10-fold in Teheran and Karachi, 9-fold in Delhi and São Paulo, and 8-fold in Lima and Mexico City.

The most striking fact shown in Table 2.1, however, is that growth in most mega-cities is slowing; only Dhaka, Calcutta, and Bombay are growing faster than in the early 1950s, and all three were affected in the earlier period by the postindependence turmoil that rocked South Asia. This raises an interesting question. What accounts for this universal deceleration? A partial explanation may be the UN projection method used for the 1980-1985 growth rates because it builds in deceleration tendencies; but most of the cities show deceleration even in 1970-1975, when growth rates were based on real rather than projected data. Another possible explanation is that the policy measures adopted by almost all countries to slow big-city growth were beginning to take effect. But this notion is unconvincing because the direct policy measures to influence primacy were very weak (Richardson 1987a) and swamped by powerful

Table 2.1 Growth of the 30 Largest Cities, 1950-2000

	Population (thousands)			*Annual Growth Rate (%)*		
Mega-City	*1950*	*1975*	*2000*	*1950-1955*	*1970-1975*	*1980-1985*
Mexico City	3,050	11,610	25,820	5.26	4.81	3.58
Tokyo	6,736	17,668	24,172	4.86	1.52	1.26
São Paulo	2,760	10,290	23,970	5.73	4.50	4.28
Calcutta	4,520	8,250	16,530	2.05	2.96	2.76
Bombay	2,950	7,170	16,000	3.29	3.63	3.32
New York	12,410	15,940	15,780	1.36	−0.43	0.04
Seoul	1,113	6,950	13,770	6.67	4.99	3.88
Rio de Janeiro	3,480	8,150	13,260	3.86	2.55	2.37
Shanghai	10,240	11,590	13,260	0.38	0.32	0.36
Jakarta	1,820	5,530	13,250	4.14	4.20	3.52
Delhi	1,410	4,630	13,240	4.89	4.84	4.61
Buenos Aires	5,251	9,290	13,180	2.91	1.68	1.56
Karachi	1,040	4,030	12,000	5.48	4.98	5.20
Teheran	1,126	4,267	11,329	6.52	5.76	5.33
Beijing	6,740	8,910	11,170	0.95	1.44	0.43
Dhaka	430	2,350	11,160	4.02	8.42	7.25
Cairo	3,500	6,250	11,130	2.46	1.87	2.15
Baghdad	579	3,830	111,25	4.35	5.10	4.23
Osaka	3,828	8,649	11,109	4.22	0.88	1.62
Manila	1,570	5,040	11,070	3.83	6.70	3.30
Los Angeles	4,070	8,960	10,990	4.83	1.22	1.05
Bangkok	1,440	4,050	10,710	4.18	4.30	4.05
London	10,369	10,310	10,510	0.19	−0.54	0.10
Moscow	4,841	7,600	10,400	2.68	1.44	1.79
Tianjin	5,450	7,430	9,700	1.07	1.56	0.63
Lima	1,050	3,700	9,140	5.01	4.71	4.25
Paris	5,525	8,620	8,720	2.59	0.66	0.04
Lagos	360	2,100	8,340	6.60	7.51	5.34
Milan	3,637	6,150	8,150	2.05	2.18	1.41
Madras	1,420	3,770	8,150	1.92	3.81	3.11

SOURCE: UNDIESA 1987d, 142-143, 148-260.

implicit spatial (macro and sectoral) policies that reinforced big-city growth. A third possible explanation arises from shifts in the international structure of relative prices for primary products, energy, and manufactured goods after 1973 that made the growth environment for many developing countries much less favorable (Kelley and Williamson 1984). This may have been a factor in some cases, because urban growth reflects economic growth, but in some countries mega-city growth had already begun to slow before 1973, whereas others prospered quite well in the post-1973 environment. A fourth possibility is that market forces spontaneously slowed mega-city growth, perhaps reflecting declining productivity advantages and a commensurate increase in the locational attractiveness of secondary cities (indeed, in a few countries, e.g., Korea and Peru, secondary city growth rates have begun to move ahead of those of the primate city) or a perceived deterioration in the relative positive consumption externalities of large cities relative to small. But the incidence of such a market trend was too spotty to explain the general pattern of mega-city deceleration. A more convincing and more general explanation is that the slowing of mega-city growth simply reflects national demographic trends—a deceleration in the growth rate of national population as a whole. The only exceptions were Nigeria and Egypt, although deceleration was very modest in a few other cases (Bangladesh, India, Pakistan, Indonesia). As yet, the slowing in mega-city growth rates appears to be much more a demographic than an economic phenomenon.

The Economic Efficiency of Mega-Cities

A key question is whether mega-cities have efficiency advantages for the location of economic activities. Much of the emphasis on agglomeration economies in the literature of the past 15 years suggests that they have, but the empirical basis for such a conclusion remains quite weak. Measuring agglomeration economies has been only partially successful, despite some ingenious efforts. Much of the research has focused on estimating economy-of-scale parameters over a range of city size classes in which the mega-city has been one observation or a missing variable; extrapolation from results from smaller size class ranges is very dangerous. Most of the detailed studies have been on the United States or other developed countries, such as Japan. The number of developing country applications can almost be counted on one hand: Brazil (Henderson 1986, 1988; Richardson 1984a; Rocca 1970); India (Shukla 1984); Korea (Hong 1987); and Pakistan (Qutub and Richardson 1986).

Although the number of research studies on the economic advantages of city size in developing countries is too small to draw conclusive results, the evidence is less supportive of the strong agglomeration economies argument than in research on developed country cities. The most detailed research so far is Henderson's (1986, 1988), primarily on

Brazil, with more general observations about China. Henderson's key finding was that *localization economies* (external economies associated with spatial clustering in the same industry) are much more important than *urbanization economies* (agglomeration advantages for all, or at least many, economic activities), and that localization economies are stronger for heavy industry than for light industry. Also, localization economies had declining elasticities—they became weaker with increasing urban scale. Because the exploitation of localization economies implies city specialization, Henderson (1988, 97) concludes: "The issue then is not whether bigger cities are more productive than smaller—in net terms they are equally efficient. The issue is what types of industries are better off in small cities vs. bigger cities."

Nevertheless, an implication of Henderson's findings, underscored by the high concentration of all types of manufacturing in China's three largest cities (Shanghai, Beijing, Tianjin), is that it may be inefficient to focus industrial development in mega-cities, especially for many heavy industries and for industries producing standardized goods. At least three arguments support this view. First, as stressed in Henderson's research, many industries benefiting from strong localization economies can exploit their scale effects in small or medium-sized cities, where land and labor costs are lower (a good Brazilian example is the shoe industry in Franca in the northern fringe of São Paulo state). Second, it is usually more efficient to locate materials-oriented, weight-reducing production activities, such as iron and steel, close to resource deposits rather than near major consumption markets (e.g., in mega-city locations). Third, many heavy industries (again, iron and steel is a good example, along with chemicals, cement, and petroleum refining) are heavy polluters, so a mega-city location would maximize the pollution damage.

Other findings for Brazil are consistent with Henderson's conclusions. Rocca (1970) estimated output elasticities for 1960 manufacturing somewhat higher than Henderson's 1970 estimates, indicating positive agglomeration effects but not strong enough to suggest ever-increasing scale economies. Richardson (1984c) found that many medium-sized cities ranked much higher than São Paulo (ranked 25th) in industrial value added per capita.

But Henderson did find that big cities have some advantages that could make them attractive to certain types of industry. For example, the large metropolitan areas in Brazil had a favorable labor skill mix (a high ratio of skilled to unskilled workers), although he explained this much more by consumption externalities (e.g., the greater pull of larger cities because of superior education and health facilities and a wider range of consumer goods and services) than productivity advantages. Shukla (1984) supports this argument in her analysis of Indian manufacturing agglomeration economies. Urbanization economies were much more important than localization economies, possibly suggesting the importance of the general availability of services and skilled labor in the larger Indian cities rather than the specialized local labor pools that seem so critical to Brazil's more sophisticated industrial structure.

The other relevant research relates to Korea and Pakistan. Using a Dhrymes-type indirect (wage) production function to measure returns to scale in Korean manufacturing, Hong (1987) found modest signs of agglomeration economies, with scale parameters not high enough to suggest that Seoul (or Pusan) had overwhelming efficiency advantages. Qutub and Richardson (1986) examined how nonagricultural value added per urban employee varied across 293 urban areas larger than 10,000 people in Pakistan. This measure is far from perfect, because interurban variations in value added could reflect industry-mix effects. Nevertheless, subject to this qualification, the analysis is suggestive. Although Karachi's productivity was 53 percent higher than the national urban average, it was lower than in five district capitals (all with under 250,000 population). Moreover, the relationship between productivity and city size was very blurred. The best performers tended to be small or medium-sized specialized manufacturing towns, consistent with Henderson's emphasis on localization economies in Brazil. Finally, Lahore—not quite a future mega-city but heading for the 6-million-plus class—is one of Pakistan's least productive cities with a productivity level almost two-fifths below the national urban average. Incidentally, Lahore was the location of the World Bank's major urban project in Pakistan just when the bank was stressing the need to invest in the most efficient cities! This example reminds us that the World Bank cannot plead total innocence for simplistically equating urban economic efficiency with very large cities in the late 1970s and early 1980s.

A possible challenge to these results about the moderate degree of city-size-related agglomeration economies is the empirical evidence on the awesome concentration of economic activities relative to, say, population share in mega-cities. It would be wrong to explain this concentration solely in terms of economic efficiency, however. The strongest case for such an explanation can likely be made with respect to high-order service industries, especially information processing, but these industries have not received much attention in the research on developing country agglomeration economies, mainly because of severe data limitations. Spatial concentrations of manufacturing in mega-cities, however, may reflect less inherent productivity advantages than the intended and unintended effects of government policies. Import substitution strategies have reinforced the locational pull of the dominant national market (the primate city) on mobile industries. Food, energy, transport, and utility subsidies have kept labor costs in the largest cities unnaturally low, as they would have by far the highest costs if marginal cost-pricing rules prevailed. The widespread use of *discretionary* industrial subsidy, tariff protection, and capital subsidy schemes has made access to government a major factor in industrial profitability in those mega-cities that are also government capitals. Entrepreneurs have frequently showed great reluctance to evaluate up-country locations because of personal locational preferences and mega-city consumption externalities for the elite (private schools, universities, hospitals, cosmopolitan cultural activities, restaurants, etc.). Finally, the countries with mega-cities are scattered all over the development spectrum, and in the least developed countries the spatial concentration of modern economic activities in the primate city is the natural, but temporary, consequence of inadequate demand thresholds outside the core region. As development occurs, the locational advantages of the mega-city may progressively weaken.

The Capital Cost Burden

Mega-cities are probably much more expensive than smaller cities in capital costs of infrastructure and services. The service needs of large cities, high densities, population pressure, and the status considerations associated with a national capital are some of the relevant considerations. In addition, the higher incomes earned in mega-cities lead to demands for very high standards of services relative to other cities. Construction costs, especially labor costs, are usually much higher than elsewhere. Land values are an exponential function of city size. Even if land costs are not a resource input but a transfer item from firms and households to landowners, they are relevant in that they directly strike at affordability and they may drain government revenues if the state is required to compensate for compulsorily acquired land.

In the 1980s the capital costs of urbanization were measured for four mega-cities (Cairo, Jakarta, Karachi, Dhaka) relative to smaller cities in their respective national urban hierarchies. In three, per capita housing, urban infrastructure, and job creation costs were far higher (Richardson 1987b); the exception is Cairo, largely explained by assumptions underpinning the research rather than by objective reality (PADCO 1982). Probably mega-cities consistently suffer from higher per capita capital costs of population absorption than other urban places. In addition, the international prestige governments believe their largest cities generate as national showcases leads to capital expenditures for cultural centers, five-star hotels, government buildings, and other architectural landmarks. In any event, the capital cost issue is much more critical than the familiar argument that managing large cities incurs higher operating costs. In fact, almost all developing countries suffer, more or less, from capital constraints, and urban absorption costs mop up a large share of the national investment resource pool. Costs of providing housing and basic urban services to their mega-city populations over the next two decades could absorb 8-18 percent of the total available resource pool in these four countries (Richardson 1987b).

I could argue that the mega-city capital cost burden is such a serious issue that asking whether it could be alleviated is a key policy question. Certainly, if would-be migrants could be kept "down on the farm," the capital cost savings could be dramatic, because mega-city absorption costs per capita are four to six times greater than rural absorption costs. But attempts to stem rural-urban migration have been a losing game in developing countries, although there may be some interventions at the margin that might help, such as rural public works programs (e.g., farm-to-market roads, small or medium-sized irrigation projects). Moreover, even if rural retention policies could slow rural to mega-city migration, any capital cost savings might be offset somewhat by GNP losses. Attempts to divert migrants to smaller cities might seem a more attractive proposition, but the savings here are largely illusory. The problem is that lower capital costs of smaller cities are partly offset by higher capital costs in the peripheral regions where many of them are located. The result is that capital cost savings of a more decentralized urban settlement pattern turn out to be far more modest than expected, probably 4-7 percent of total urban capital costs (Richardson 1987c).

Could the higher mega-city output generate a flow of investment resources (out of output) to pay for the extra costs of absorbing workers and their families in the mega-city compared to the rural area? Consider a hypothetical case. If marginal mega-city absorption costs per capita were $4000 and rural absorption costs were $1000 per capita, and assuming labor force participation at 40 percent, the net increase in marginal product per worker associated with migration would have to generate enough investment resources to bridge a gap of $7500. To keep the analysis simple, assume that the urban capital stock lasts forever without depreciation and maintenance (this underestimates costs). With a 10 percent discount rate and a gross domestic investment/gross domestic product ratio of 25 percent, the marginal product differential between the mega-city and the rural area would have to be $3000 ($7500 × 0.10/0.25)—higher than the prevailing gap in the vast majority of developing countries.

This conclusion suggests a bitter choice: either diversion of an ample share of scarce investment resources to mega-city population absorption or a chronic inability to provide housing and services to its expanding population. Even worse, even in those countries with the investment resources available to finance the growth of the mega-city, it may be very difficult to mobilize the needed resources from a budgetary point of view. In most countries, the bulk of urban infrastructure is provided by the public sector, the urban development share in national development budgets is usually small, and it may not be feasible in political terms to shift funds among sectors fast enough to cope with the mega-city investment crunch. The only other plausible solution, although it is often very effective, is to cut mega-city infrastructure and service standards. In many developing countries, standards are so high that cuts of 50-60 percent are possible if the reductions can be implemented institutionally (Gakenheimer and Brando 1987). Capital cost savings benefits are supplemented by increased affordability, making it easier for a higher proportion of mega-city households to pay a bigger share of service provision costs. Thus, reductions in infrastructure standards make cost-recovery programs more feasible, in turn helping to reduce service deficits.

Social Benefits

In a provocative 1983 paper, Mera and Shishido argue that developing countries with primate city size distributions scored much better on social indicators, such as nutrition, infant mortality, and literacy, than countries with less skewed national urban settlement patterns. Although their research was intended as a social analogue to Mera's (1973) earlier finding of the association between economic development and primacy, and their hypothesis has not to my knowledge been subjected to rigorous empirical testing by other researchers, the argument suggests potential social benefits in mega-cities. The development literature is riddled with ad hoc statistics on interurban and interregional differentials in input measures of social output (e.g., pupil-teacher ratios, population per doctor or nurse, number of health centers per capita) that tend to support the conclusion that social indicator scores are much higher in the largest cities than in smaller urban places.

This uneven distribution of public output is paralleled by a similarly uneven distribution of private goods and services, especially in culture, entertainment, and leisure. In economics jargon, the spatial concentration of private and publicly supplied goods in mega-cities is a consumption externality, possibly as important as the production externalities that account for the disproportionate concentration of economic activities in these cities. But the influence of these consumption externalities has to be analyzed with some caution. Especially in the lower-income mega-cities, high social indicator scores may not have much impact on their attraction to migrants, mainly because there the consumption of health, education, and other social services is so uneven, geographically and by social and income class. Wealthy neighborhoods not only have the benefit of private schools and hospitals but also often receive a disproportionate share of publicly supplied services. The paucity of both in smaller urban centers makes them very unattractive as potential locations for big-city elites—a major obstacle to the decentralization of economic activity and public administration. In middle-income countries where social services are more equally distributed, this problem may be less serious. Decentralizing educational institutions out of Seoul, for example, has helped to make some other cities more appealing for mobile firms (and their managers) in recent years.

It is sometimes argued, with justification, that migrants in developing countries move for economic reasons. But to conclude that the spatial distribution of social facilities has no influence on the spatial distribution of economic activity would be wrong, as it ignores the resistance of mega-city elites to dispersion because the smaller cities too frequently lack the range and quality of social and personal services these persons demand. So the pace of decentralization of economic activity is much slower than if these services were more equally distributed in national urban space, and mega-city growth has been commensurately faster. This theory may help to explain why policymakers have shown far more interest in reducing *interurban* differentials in social indicators than in narrowing the very wide *intraurban* differentials.

Mega-City Disamenities

Mega-cities in developing countries suffer to varying degrees from negative externalities such as poor air and water quality, chronic traffic congestion, inadequate solid waste disposal, sewerage deficiencies, and in a few cases, high crime rates. These disamenities are often given more attention than the agglomeration economies and other efficiency advantages, suggesting the concept of the "big, bad city" (Richardson 1989a). Almost all the hard data on urban disamenities, however, come from developed countries, so conclusions about developing countries are based on inferences from research in the developed world, reinforced with scraps of information about individual Third World cities and visitors' personal observations and experiences. The lack of comparative research makes it difficult to quantify the extent of the negative externalities problem in developing country mega-cities, and this deficiency undermines conclusions about their aggregate welfare effects.

Information about negative externalities in Third World mega-cities is spotty, if not anecdotal, but a few generalizations are possible. First, some problems are more serious in some cities than in others. Air pollution is very serious in Mexico City, São Paulo, Bombay, Manila, Calcutta, and Seoul, whereas water quality is a major problem in Dhaka, Delhi, Seoul, Karachi, Bangkok, and Manila. Traffic congestion is particularly serious in Bangkok, Teheran, Mexico City, Cairo, Manila, and Lagos, and overcrowded public transport facilities are especially bad in the Indian mega-cities. Crime rates are particularly high in Lagos, Lima, Rio de Janeiro, and São Paulo. Sewerage problems are acute in Bangkok, Manila, Dhaka, and Karachi, and solid waste collection is grossly inadequate in Lagos, Jakarta, Lima, and Karachi.

Second, consequences of negative externalities do not appear to elicit the same kind of policy response in developing country mega-cities as they would in developed countries. For example, reports that mortality is 250 per 100,000 from respiratory disease in Bombay, or that Delhi's Yamuna River collects 200 million liters of untreated sewage daily, or that 97 percent of drinking water samples in Madras show signs of fecal contamination pass without notice (Centre for Science and Environment, India 1989). Whereas air quality in U.S. cities has been improving, in many developed country cities it has been deteriorating; in Mexico City it has been getting worse at a rate of 3.4 percent per year (Schteingart 1989). Manila's 22.6-hectare garbage dump, known as Smokey Mountain, sustains a population of 20,000, and there has been no serious attempt to deal with this garbage problem in the last 40 years (Jimenez and Velasquez 1989). Only 2 percent of Bangkok's population is connected to city sewers, and dissolved oxygen levels in the Chao Phraya river have shown depressed sag curves for nearly 20 years without much remedial action being taken (Phantumvanit and Liengcharernsit 1989). Reasons for the limited action are not hard to find: Environmental quality is an income-elastic good that is not a high priority for poor, or even middle-income, nations.

We lack enough information to determine whether negative externalities are more serious in developing country mega-cities than in developed country mega-cities. Limited evidence suggests they are more serious *physically,* but this is offset by much lower imputed monetary costs. Multiplying physical units and dollar cost per unit for each disamenity might lead to the conclusion that negative externalities account for a similar proportion of gross urban product in the typical *middle-income* country mega-city as in the typical developed country mega-city. In poorer mega-cities, much may depend on the level and pattern of economic development and life-styles. (What is the basis of the transport system? What fuel is used domestically? What is the dominant sanitation system? How is garbage collected? etc.)

Spatial Restructuring

Policymakers have given considerable attention to decentralizing population or economic activity to smaller urban areas as a strategy to alleviate the social costs associated with rapid mega-city growth. With varying degrees of vigor and political will, the vast majority of developing countries have pursued policies to promote the growth of intermediate (secondary) cities or even small towns and rural areas via migration diversion or population retention strategies. The problems with this approach are not merely that these policies are extremely difficult to implement successfully but that there is a much simpler, more feasible option for reducing congestion and other size-related economic and social costs of mega-cities, that is, via the evolution of a planned or spontaneous polycentric spatial structure. It is reasonably clear that a mega-city cannot remain efficient as a monocentric city. If most jobs continue to be concentrated in the central city core, population growth and radial extension of the urbanized area will inevitably result in progressively more severe congestion costs. Decentralizing jobs to subcenter locations in combination with the faster growth of suburban populations (in poor countries, more likely to be low-income) will relieve congestion without sacrificing the benefits of area-wide agglomeration economies. Similarly, risks of diseconomies of scale in urban service production and delivery can be minimized via the development of two (or multiple) tiers of government, with the lower tier offering more opportunities for citizen participation and government responsiveness to citizen preferences. Thus, many of the external diseconomies that erode mega-city efficiency can be avoided by changes in the metropolitan spatial and political structure toward a more polycentric pattern. In effect, a polycentric mega-city structure avoids any need for secondary cities or similar strategies, at least if these imply interurban population redistribution.

But the case for spatial restructuring implies nothing about how it might be achieved and, in particular, whether it might be brought about spontaneously or via planning intervention. In many real-world cases, problems have developed because government intervention has consciously or unconsciously slowed decentralization or has tried to promote inefficient or premature decentralized development patterns. An example is the ambitious, even utopian, strategy to build a multimillion-population new city (New Bombay) in Bombay across Thane Creek rather than providing badly needed infrastructure and services in rapidly and spontaneously growing industrial areas such as the Kalyan complex of manufacturing towns in the northeast section of the metropolitan region (Richardson 1984d). Another example is the plan adopted in Peru during the Belaunde administration (1980-1985), but later abandoned, to promote a 600-kilometer urban coastal corridor north and south of Lima, to be supported by an underused express highway. The development corridor was too long and capital costs of the required infrastructure too expensive. A much simpler and sounder strategy would have been to improve public services and transportation at the nearby spontaneous settlements of Cono Norte and Cono Sur, especially in the squatter towns of Comas and Villa El Salvador (Richardson 1984b).

Other mega-city attempts at successful decentralization have been equally flawed. Dhaka adopted a strategy to promote a northern development corridor between Tongi and Joydepur after 1981, but the pattern of growth on the ground has deviated widely from the proposed strategy. For example, the area near the new Kumitola airport was planned as a low-income residential area but is in fact developing as a location for high-income commuters. Calcutta and Madras have pursued ambitious decentralization strategies in the past that failed to consider the cities' relative economic stagnation. Policies in Madras (relocating markets and transportation terminals to three subcenters; promoting residential development in the industrial suburb of Manali, 20 kilometers from the downtown; and building a new town at MM-Nagar, 45 kilometers south of the city) have been very slow in implementation. Calcutta has changed horses several times in recent decades, with the failed northern countermagnet (Kalyani-Bansberia) of the 1960s and the excessive number (26) of recommended subcenters after 1976, subsequently cut back to 17 in 1981. There has been some success recently with the focus on a few close-in subcenters such as Dum-Dum and Salt Lake. Cairo and Jakarta have tried to alter the natural axis of growth (from north-south to east-west), but with little success. Even Delhi, with its rapid growth fueled by a disproportionately high share of India's capital expenditures for urban development, has experienced difficulties in implementing a decentralization strategy of 18 regional subcenters (including 6 satellite towns). Seoul, however, is generally regarded as a success story. A broad mix of policy actions (e.g., new towns and satellite cities, industrial location controls, new industrial estates, huge investments in public transit, land use controls, residential taxes differentiated by city size, and decentralization of higher education facilities) has been used to promote substantial development south of the Han River (Richardson and Hwang 1987). But there has been criticism, too, over the land and house price inflation boosted by the large greenbelt (Mills and Kim 1986) and the challenging argument that industrial decentralization was the product of market forces rather than the battery of government policies (World Bank 1984).

In a few developing country mega-cities, such as Bangkok and Manila, decentralization has been predominantly a market process because no convincing decentralization strategy was in place, and the results have been quite satisfactory. In Bangkok, development has occurred most rapidly along key highways to the east (toward the industrial town of Chonburi) and north (near Dom Muang airport). There has been substantial decentralization of population and employment to five surrounding changwat (provincial) capitals—Samut Prakarn, Pathum Thani, Nontha Buri, Nahon Pathom, and Samut Sakorn, but more as a consequence of market forces than of any explicit spatial strategy. Also, the private sector in Bangkok has been successful in producing moderately priced suburban housing even as the Bangkok Housing Authority has been immobilized by high production costs and badly chosen sites (Dowall

1989). Subcenters in Manila, first in Quezon City, later in Makati, were more spontaneous than planned and have turned out reasonably well. Although Karachi has lacked a spatial decentralization strategy, major subcenters have formed around the new port of Bin Qasim and neighboring steel mill and at the Hub industrial estate across the border in Baluchistan, but as the result of government sectoral rather than spatial policies. (For more details on the Asian examples, see the UN monographs on population growth and policies in mega-cities, UNDIESA 1986-1989.)

Spatial restructuring is critical for mega-cities' continued successful growth. Several diseconomies of urban scale, such as pollution and traffic congestion, could be reduced by the transition from monocentricity to polycentricity without risking the erosion of agglomeration economies; many are economies of concentration not of centralization, although some are metropolitan-wide in scope. Mega-city policymakers have not been very effective in promoting or supporting polycentric development, in part because of the critical shortage of public capital resources and the lack of control mechanisms for allocating land uses, but these are not a total explanation. In some cases, results might have been better had policymakers avoided ambitious decentralization strategies and instead tried to follow spontaneous market adjustments—making infrastructure investments in support of decentralizing location decisions. Resources were frequently squandered by investing in infrastructure in the wrong places or too far ahead of prospective demand. In other cases, planners proposed and attempted to implement strategies that were far too ambitious for stagnant or slow-growing environments. In a few cities, policymakers became obsessed with revitalizing the central core and neglected opportunities presented by decentralization forces. But the public sector will continue to hold the prime responsibility for supplying urban infrastructure, regardless of circumstances. The key policy issues are to learn how, when, and where to intervene to promote the efficient polycentric structures that are critical to successful mega-city growth and to avoid harmful interventions that might inhibit the necessary spatial restructuring.

Managing Mega-Cities

Metropolitan management is one of the buzz words of urban policymaking in developing countries. The most serious problems probably occur at the two extremes of the urban scale—in small cities (where shortages of personnel and financial resources are chronic and debilitating) and in the largest cities, especially the mega-cities, where substantial managerial diseconomies within bloated bureaucracies and the lack of big government's responsiveness to neighborhood and individual preferences undermine the effectiveness of metropolitan management. Although some of the problems could be alleviated by multiple tiers of local government, governments of Third World mega-cities have not been very interested in following this route.

Despite the neglect of multiple tiers, some attention has been spent on appropriate institutional frameworks. Municipal administrations are not usually the right solution because their jurisdictions are underbounded and they are often fully preoccupied with daily budget, administration, and service provision problems. Although many mega-cities have, in name at least, metropolitan development and planning authorities (e.g., Metro Manila Commission, Bangkok Metropolitan Administration, Karachi Development Authority, Bombay Metropolitan Region Development Authority, Dhaka Improvement Trust), they are often weak institutions with little influence on mega-city growth and development. Even the more successful ones, such as the Calcutta Metropolitan Development Authority and Delhi Development Authority, operated primarily as public works agencies performing functions that could have been undertaken by either central or local rather than metropolitan government. Seoul and Jakarta have no metropolitan development authorities. In Seoul, prime responsibility has lain with a bureau in the central government's Ministry of Construction; in Jakarta, responsibility is shared among adjacent provincial governments and certain central government departments, especially the Directorate General of Housing, Building, Planning and Urban Development in the Ministry of Public Works.

There has been debate on whether metropolitan development authorities are conducive to effective big-city management (Sivaramakrishnan and Green 1986). If existing agencies are competently undertaking key metropolitan functions, they may not be necessary. For example, preparing a metropolitan capital investment budget is a critical management function. Yet in Mexico City, capital budgeting is a function of the central government programming and budget office, and similarly in Manila, the Capital Investment Folio is prepared by the Ministry of Public Works and the National Economic and Development Authority —for both cities, reasonably efficient. The soundest conclusion about mega-city institutional frameworks may be that there are enough differences among political and planning systems that there is no clear, universal institutional panacea. Metropolitan functions can be met by a metropolitan development authority or a regional or central government depending on circumstances; the only safe conclusion is that a municipal administration will not work well because of underbounding and the inability to respond, given daily pressures, to conditions of rapid population and economic growth.

Some analysts write as if city size were a management variable, arguing that policy interventions to control or at least influence mega-city size are appropriate ways of dealing with negative externalities and other mega-city problems. This argument is not very strong. It is almost always more efficient to deal with problems directly (e.g., emission fees to control air pollution, transportation demand management strategies to reduce traffic congestion) rather than indirectly by choking off demand via city-size manipulation.

A common argument in developing countries is that the efficiency of mega-cities is hampered by their frequent role as national administrative capitals. A dramatic but simple managerial solution might be to relocate the capital functions to a new city in some other region. The experience ranges from vague proposals that have led nowhere (e.g., the recurrent suggestion in Korea to set up a new capital somewhere in the middle of the country, President Garcia's idea in the late 1980s to return Peru's capital to Huancayo in the Andes); suggestions that may be implemented (e.g., relocating Argentina's capital from Buenos Aires to Viedma-Carmen in Patagonia); schemes that have been started but held up by lack of resources (e.g., six relocations in Africa, only one of which would affect a mega-city—shifting Nigeria's capital from Lagos to Abuja); and more or less completed projects, in particular, Islamabad and Brasilia. These last two cases illustrate some of the problems of capital relocation, such as the high capital costs of building a new center and the communication diseconomies between it and the mega-city former capital that hamper government and business efficiency and isolate politicians, ministers, and legislators from public opinion. Does relocating the national capital affect the pace of mega-city growth? The Brazilian experience suggests yes (Rio is now the slowest growing of Brazil's nine metropolitan areas) and the Pakistan experience suggests no (with no discernible impact on Karachi's growth).

The earlier discussion of spatial restructuring implies that mega-city management may involve some support to decentralization. Although many mega-city policymakers have intervened spatially, the consequences have rarely suggested managerial efficiency. Ambitious plans that failed because of insufficient capital resources, investments in the wrong places (e.g., badly located industrial estates), attempts to fight rather than support market forces, a frequent emphasis on core city development combined with failure to provide infrastructure at expanding decentralized sites, and impotence in controlling metropolitan land uses mostly suggest managerial failures, not successes. Sound mega-city spatial management may require more intervention in some spheres (e.g., measures to accelerate delivery of basic services to newly developed, low-income neighborhoods) but much less in others (e.g., large-scale land acquisition, preselection of subcenters, expensive new satellite towns).

Further, because managerial resources are very scarce in mega-cities (but even scarcer in smaller urban places), it is more efficient to deploy them where the degree of effective intervention needed is modest. Pricing and tax-subsidy policies may be very economical of managerial resources. A good, if politically volatile example, is eliminating food subsidies. The resulting increase in food prices would be very effective in changing the internal terms of urban-rural trade, thereby reducing migration to the mega-city. This approach is very sparing in its use of managerial resources compared with alternatives such as an investment and accommodationist strategy to ease the absorption problems of new urban migrants. The risk of a high political price suggests proceeding with caution, reducing food subsidies over several years. Another approach to economize on scarce public managerial resources is privatization as a way to manage mega-city growth. Most applications are in transportation and housing, but in many developing countries with hospitable environments for nongovernmental organizations and entrepreneurs there is considerable scope in the health, education, and social welfare sectors.

The ability to deliver basic urban services to growing mega-city populations may be an even better test of managerial effectiveness. Many mega-cities suffer from severe service deficits (frequently affecting more than half the population) because of insufficient capital resources and often a conscious decision to provide very high standards of services in well-off neighborhoods while completely neglecting poor ones. Although urban service deficits are probably more severe in small than in large cities (Indonesian evidence suggests that water, sanitation, and electricity deficits were 10, 20, and 50 percent more severe; PADCO-DACREA 1985), the large city problem may be politically more threatening because of higher expectations. Also, the nature of the managerial problem changes with city size; in a mega-city the scale of service projects necessary is a major difficulty, whereas in smaller cities the key issue is availability of resources. Also, in many countries resource scarcities are aggravated by a tendency to treat many urban services (water, sanitation, education, health, even housing) as *social* services rather than user-fee ones, making it very difficult to introduce cost-recovery schemes. Public works agencies quickly use up the financial resources to keep service provision not too far behind population growth.

The fiscal resources issue is critical. A few mega-cities have received substantial aid from international lending agencies (Calcutta's long and cozy relationship with the World Bank's IDA arm is a notable example), but chances of obtaining substantial aid from multi- or bilateral lenders or even from higher levels of government are modest, though mega-cities may, via their political influence, be able to squeeze more out of the national government than smaller, more peripheral cities. In the 1970s, metropolitan areas with over 500,000 population received about half of all multilateral aid for human settlements—well over their 15 percent population share (Blitzer, Hardoy, and Satterthwaite 1988), though shifts in policy emphasis toward smaller cities may be eroding this share. The prevailing opinion now is that cities, including mega-cities, must rely more and more on internally generated resources to finance services. But revenues from property and other taxes suffer from lax collection, infrequent adjustment in rates, and for some taxes, compulsory pass-through provisions to higher levels of government. Jurisdictional fragmentation leads to severe disparities within metropolitan regions, especially in peripheral low-income areas with low tax receipts and huge service needs. The lack of buoyancy in tax revenues combined with inadequate funding from above explain the growing appeal of the cost-recovery approach. Political objections may be loud and institutional arrangements less than ideal, but cost-recovery schemes work because households are willing and able (given innovative financing provisions) to pay for guaranteed basic services such as piped water, low-cost sanitation, and electricity, especially if the alternatives are doing without or buying more costly private substitutes.

Imperfections in operating and maintaining urban service systems are a major managerial deficiency, even in some mega-cities. The action needed is often simple—repairing leaking standpipes, replacing broken electricity and water meters, regular updating of cadastrals, keeping streets and drains clear of garbage—but these are often ignored, undermining the benefits from urban capital projects. And although the technological solutions are easy, solving managerial problems is not—

in territorially large metropolitan areas, a spatial network of engineers is needed for monitoring and upkeep. Maximizing the use of existing facilities may yield a huge payoff by relieving the strain on new capital projects, for example, improving traffic management schemes (Bombay, Bangkok, Lima), managing water leakages (Manila, Karachi, Jakarta), and upgrading slums as a substitute for new housing (Rio, Calcutta, Jakarta).

Although developing country mega-cities have not been able to expand services fast enough to keep pace with population growth, resulting in widening service deficits, and they have been hampered by capital resource limitations and inadequate cost recovery, nevertheless many of them have made considerable progress. Have their managers performed worse than their developed country counterparts, especially those in older cities needing major renovation of aging infrastructure? Is New York doing any better than São Paulo, Jakarta, or even Cairo or Calcutta? Managers in the older cities of developed countries and the mega-cities of the least developed countries may find it most difficult to cope, whereas those in countries at an intermediate stage in development, with the twin benefits of a slower pace of urbanization and relatively slack resource constraints (e.g., in Seoul), have an easier time.

Policy Options

Sensible options for mega-city policymakers are implicit in much of the preceding analysis. The most important conclusion, if a negative one, is that mega-city size is not a critical policy variable. Rammed home by many analysts over the years (perhaps most forcibly by Mills in Cameron and Wingo 1973 and Mills and Becker 1986), many policymakers still ignore this point. The size of a mega-city is not closely correlated with its economic efficiency or the severity of its negative externalities, and attempts to change its size—either absolutely or relatively—would, even if successful, have no direct impact on economic performance or disamenities. Even the growth rate of the mega-city need not be a problem unless it is so high that it impedes economic growth, fiscal stability, provision of public services, and implementation capacity. Such growth management costs are triggered at different growth rates in different cities, within and across countries, but (see Table 2.1) no mega-city appears to be experiencing growth rates in the dangerous range. Effective mega-city *management* is much more critical than mega-city *size*, if that management is defined not narrowly in terms of finance and service supply, but more broadly to include coordination of macro-, sectoral, and mega-city policies; strategies to deal with negative externalities directly (e.g., measures to control pollution, transportation demand, and system management); facilitation of efficient spatial reorganization as the mega-city continues to grow; and actions to improve service distribution among neighborhoods and income groups to minimize the risks of social unrest and political stability. These tasks are far from easy to perform successfully, but they have the virtue of relevance, which is more than can be said for policies to control mega-city size. Among these, one of the most dangerous is the "growth cap" strategy that aims to cap future mega-city growth at some predetermined limit. This is a recipe for policy failure. It runs the risk of underinvesting in mega-city infrastructure, resulting in capital cost savings being swamped by higher losses because of operating costs and social costs (e.g., congestion). To the extent that such a strategy was partially successful, the consequences could include intolerably high growth rates for smaller cities and aggravation of the problems associated with excess rural labor.

The search for effective mega-city size control measures is likely to be futile. In-migration controls are enforceable only via draconian rustication programs (UNDIESA 1981). Restrictions on public investment, by slowing the pace of new investment or by deliberate undermaintenance of the mega-city capital stock, may undermine *national* economic growth and permanently deprive poor neighborhoods of access to basic urban services. Controls on mega-city formal-sector employment and investment growth are difficult to implement politically and have hidden costs in abandoned projects or economic activities diverted to higher-cost locales. The most favored approach is implementation of policies to change the distribution of population and economic activity in the national economy, including programs to hold population in rural areas; promoting intermediate (secondary) cities or small towns; investing in border regions; experimenting with land colonization schemes; developing new towns; designating growth centers, countermagnets, or other priority nodes; decentralizing the core region to satellite cities or district capitals; and relocating the national capital. These policies can be very expensive, they are prone to failure, and their impact on mega-city growth is usually minimal (Richardson 1985).

Accommodationist measures (Laquian 1981, 101-112) in the mega-city may tend to slow growth, but for a perverse reason. These measures include legalization of squatter tenure rights, slum upgrading, guided land development and residential sites and services provision, promoting informal-sector activities, and loans for microbusinesses. Sometimes seen as stimuli to mega-city growth, in practice these measures tend to have the opposite effect. Most of the measures (except interventions to promote economic activity among the poor) involve diverting scarce investment resources from economic to social infrastructure. There is no evidence that accelerating urban service provision to poor mega-city residents in developing countries (the key focus of the accommodationist approach) results in higher in-migration.

The other form of intervention that could affect relative size of the mega-city without socially damaging effects consists of policies to reduce, if not eliminate, distortions in the real income differential between the mega-city and the rest of the country, especially rural areas. In most developing countries, this differential is widened by subsidies to the mega-city for food, mass transit, utilities, and by unintended side effects of other macropolicies and sectorally implicit spatial policies (e.g., industrial tariffs). If all mega-city subsidies were eliminated, the differential between mega-city incomes and those elsewhere in the country would narrow, and mega-city in-migration would slow. If developing country policymakers are serious about population distribution goals, why do these subsidies remain? The explanation is simple. Especially in the vast majority of cases where the mega-city is also the national capital, policymakers are very sensitive to the more vocal needs of the mega-city populace, and there have been too many nasty experiences (e.g., food riots) indicating what is likely to happen if food and other subsidies are cut, particularly if the actions are precipitously taken.

Research Needs and Priorities

There has been considerable academic and practical discussion about levels and patterns of urbanization for at least three decades. The pros and cons of primacy have been debated ad nauseam, even if the focus on the very large city (the mega-city) is more recent, simply because there were so few cities in this size class before 1970. In addition, there has been substantial research on the relationships between economic efficiency and disamenities and city size in developed countries, especially in the United States and Japan. For these reasons, the research problem is not so much a lag in conceptualization or hypothesis formation as it is a dearth of detailed and sound empirical studies on developing country mega-cities—individual case studies and comparative research within and across subcontinents. The major obstacles to the growth of research findings are data, data, and data.

The problem is not the total absence of empirical information but rather that what information has been assembled is so scattered, unsystematic, often outdated, and almost always noncomparable. Different researchers stress different data priorities, so my suggestions are merely illustrative rather than comprehensive. One valuable step would be for national statistical agencies to cooperate to develop internationally standardized ways of collecting urban data to make international comparisons feasible. For example, Standard Industrial Classification (SIC) codes are broadly comparable across countries, and it might be feasible to develop some kind of Standard Metropolitan Classification (SMC). The UN projections (Table 2.1) reflect the problems involved in obtaining even metropolitan population estimates. Ideally, we need a "flexible boundary" approach that tracks the *urbanized* population of an ever-expanding mega-city region, but even in the United States it was not until 1980 that the Census Bureau began to present a significant amount of its data according to urbanized area rather than standard metropolitan statistical areas (SMSAs, now MSAs, that are aggregations of whole counties). It is doubtful if low-income developing countries have the resources and trained personnel to gather data on the rapidly changing urbanized area when they have difficulty keeping their cadastral surveys needed for property tax collection up to date. Of course, if all information is collected at a very fine spatial grain, one can use spatial aggregation techniques to build any combination of geographical zones needed. This technique would be very useful not only for urbanized area measures but for the intrametropolitan spatial analyses so critical in understanding the process of mega-city transformation, but costs and limited technical capacity are major constraints.

As for population data, the typical 10-year census interval is a particular problem in the analysis of mega-cities where the metropolitan population might easily grow by more than 2 million within a 5-year period. Only the greediest demographer might try to make a case for more frequent *national* censuses, but it may be feasible within the mega-city alone to mount annual surveys, probably stressing migration behavior, in recently established or rapidly growing neighborhoods. As the need here is to avoid ad hoc noncomparable surveys, standardization by location, time, and country is critical. To the extent that such surveys rely on academic research groups rather than census bureaus, the best that can be hoped for is a common core element to the survey, because each researcher is likely to have specific research interests. Similarly, conditions and behavior vary across countries (e.g., in some countries self-help housing is common, whereas in others it is unknown; in some countries men dominate migration, and in others women and families are important)—another reason why surveys would need to vary.

It is quite surprising that information about mega-city economic structure is so thin even in developing countries with excellent national economic data. The problem is that few countries undertake regular and geographically disaggregated economic censuses. Also, the metropolitan boundary problem undermines serious attempts to measure economic decentralization by distinguishing between core city and suburban or exurban economic activity. The important role of the informal sector in most developing country mega-cities increases the need for an international definition, for a method of reconciling this definition with the way statistics are recorded in individual countries, and for new ideas on how to estimate the components of this sector that slip through any official data-gathering system. A few ad hoc studies have produced a mine of information about industrial development and location in specific mega-cities, such as the studies sponsored by the World Bank of São Paulo (Hamer 1983; Townroe 1983) and Seoul (Lee 1985), but three characteristics (substantial financial support from an international agency, a strong international and national research team, and a major primary survey component) make replication elsewhere difficult.

Another important data need is to assemble internationally comparable data on the negative externalities of mega-cities—air and water pollution, traffic congestion, crime rates. Many cities have sketchy and intermittent information insufficient for more than casual empirical observation. Standardizing measures is important (e.g., for ozone levels, do we want information about peak 1-hour emissions or x-hour averages?), but a more intractable obstacle is the lack of statistical infrastructure (e.g., insufficient monitoring stations for air and water quality; zero or infrequent traffic counts; deficiencies in crime-reporting systems). Assembling the needed information on disamenities regularly calls for financial and human resources that many developing countries either lack altogether or put to higher-priority uses. Yet without such information it is difficult to draw conclusions about the welfare effects of mega-city life.

Because a major managerial problem facing mega-cities is mobilizing the financial resources needed to supply urban services, better data are needed on local budgets. Most mega-cities contain individual municipalities rather than a single metropolitan financial authority, and obtaining up-to-date revenue and expenditure information for each municipality within the mega-city region is often difficult. Moreover, without internationally accepted conventions on constructing municipal financial accounts, comparisons across mega-cities are hard to make. Without improved data on local public finances, it is impossible to evaluate mega-city performance in the financing of urban service delivery.

Most of these observations on mega-city data needs are made with the desirability of good information for comparative research in mind, but the importance of detailed case studies on individual cities should not be overlooked. For example, the World Bank's Bogota City Study (a city just outside the mega-city size class; Mohan 1979) illustrates how a detailed assessment of metropolitan issues and problems can add to understanding and policy diagnosis. Harris's (1978) study of Bombay is also useful, though slightly skimpy in coverage; Richardson's follow-up on Bombay (1980) remains fugitive. Other case studies, such as the useful and on-going United Nations series (UNDIESA 1986-1989) with their advantage of a common format and coverage of issues, and the case studies in the second volume of Dogan and Kasarda (1988b) are also helpful, but a little too brief. Monographs on individual cities that cover their historical development and analyze demographic, economic, social, and political characteristics and trends and major policy issues would add greatly to our knowledge about mega-cities.

3

Urbanization, Development, and Economic Efficiency

A. S. OBERAI

The trend in urban concentration and the growth of large cities poses a major challenge to developing countries. By the year 2000, over 45 percent of the 5.1 billion population of the developing countries will be living in urban areas, and more than 49 Third World cities are expected to have a population of over 4 million. Most population distribution policies designed to moderate the rate of rural-urban migration appear to have had limited success so far (see Oberai 1987 for a detailed assessment of migration-influencing policies). Even if such policies do succeed in the future, large cities are likely to grow larger because of the high rate of natural population increase in urban areas. In 1985, of the world's 20 largest cities, 13 were in the less developed regions. By the year 2000, 17 of the 20 largest cities will be in the less developed countries, and of the 7 super-cities with populations of 15 million or more, 5 will be in developing countries; 2 in Latin America will have a population of more than 22 million (Table 3.1).

The implications of these demographic trends for protecting the urban environment, creating employment, and providing food, housing, social services, and infrastructure are staggering. In most large cities in developing countries, at least 25 percent of the inhabitants live in absolute poverty; recent migrants and women are particularly underprivileged in terms of job access and incomes. Congestion, pollution, lack of appropriate shelter, infectious disease, and rising crime rates

Table 3.1 Largest Urban Agglomerations by Population Size

Agglomeration/ Country	*Population in 1980 (millions)*	*Agglomeration/ Country*	*Population in 1990 (millions)*	*Agglomeration/ Country*	*Population in 2000 (millions)*
Tokyo, Japan	16.9	Mexico City, Mexico	20.2	Mexico City, Mexico	25.6
New York, USA	15.6	Tokyo, Japan	18.1	São Paulo, Brazil	22.1
Mexico City, Mexico	14.5	São Paulo, Brazil	17.4	Tokyo, Japan	19.0
São Paulo, Brazil	12.1	New York, USA	16.2	Shanghai, China	17.0
Shanghai, China	11.7	Shanghai, China	13.4	New York, USA	16.8
Buenos Aires, Argentina	9.9	Los Angeles, USA	11.9	Calcutta, India	15.7
Los Angeles, USA	9.5	Calcutta, India	11.8	Bombay, India	15.4
Calcutta, India	9.0	Buenos Aires, Argentina	11.5	Beijing, China	14.0
Beijing, China	9.0	Bombay, India	11.2	Los Angeles, USA	13.9
Rio de Janeiro, Brazil	8.8	Seoul, Rep. of Korea	11.0	Jakarta, Indonesia	13.7
Paris, France	8.5	Beijing, China	10.8	Delhi, India	13.2
Osaka, Japan	8.3	Rio de Janeiro, Brazil	10.7	Buenos Aires, Argentina	12.9
Seoul, Rep. of Korea	8.3	Tianjin, China	9.4	Lagos, Nigeria	12.9
Moscow, USSR	8.2	Jakarta, Indonesia	9.3	Tianjin, China	12.7
Bombay, India	8.1	Cairo, Egypt	9.0	Seoul, Rep. of Korea	12.7
London, UK	7.7	Moscow, USSR	8.8	Rio de Janeiro, Brazil	12.5
Tianjin, China	7.3	Delhi, India	8.8	Dhaka, Bangladesh	12.2
Chicago, USA	6.0	Osaka, Japan	8.5	Cairo, Egypt	11.8
Jakarta, Indonesia	6.0	Paris, France	8.5	Metro Manila, Philippines	11.8
		Metro Manila, Philippines	8.5	Karachi, Pakistan	11.7

SOURCE: UNDIESA 1991, Table A.10.

add to misery indexes frequently highlighted by the media and assistance agencies.

Despite these real and growing problems, there is another, more positive, side to the issue, which I address in this chapter. Cities often play a critical role in the development trajectory of Third World nations. Historically, industrialization and economic development have been associated with considerable migration to the expanding centers of labor demand. To make the best use of scarce public-sector resources by capitalizing on economies of scale, facilities for power generation and water treatment, transportation systems, and other public infrastructure tend to be located in urban areas. Access to such urban infrastructure confers a cost advantage on industrial firms located in those areas, as do the economies of scale associated with access to larger and more diversified markets for labor and other input factors.

In the development scenario economists usually postulate, the increase in industrial output leads to relatively more high-wage industrial employment. Savings tend to increase as a result of consequent improvement in incomes, providing funds for investment in industrial capital. As incomes increase, the composition of domestic demand also tends to shift from food to nonfood goods, including modern health care, housing, and manufactured goods, thus further stimulating modern-sector growth. With continued changes in the composition of output and consumption patterns, the agricultural share of employment declines as the share of industry and services increases (Kuznets 1966; Oberai 1978).

Manufacturing plays a vital role in economic development. Because the income elasticity of demand for manufactured goods is considerably greater than that for food and agricultural products, manufacturing can be expected to grow relatively faster. Not only does it grow more quickly than other sectors, but its growth is normally associated with increased employment (Singh 1989). The expansion of manufacturing also increases the pace of technical change, thus helping to raise the overall rate of productivity growth in the economy.

In response to expanding modern-sector employment, migration from rural areas in search of higher incomes continues. As urban population densities increase, however, the price of urban land rises, driving up the cost of housing and other urban amenities. This narrows the real rural-urban income gap, thus slowing migration and the pace of urbanization. As development proceeds, modern-sector economic activity diversifies, and the urban sector diffuses into an integrated system of cities, each tending toward specialization in some set of economic activities.

In this description, urbanization contributes to overall development by attracting human resources to activities with greater economic returns. The movement of labor from relatively low-income, rural activity to higher-income, industrial, and modern-service sectors contributes to higher overall average income levels, further stimulating economic growth.

Real-world conditions do not always conform to the hypothetical framework of the economists' development scenario. Although contemporary urbanization rates in developing countries are comparable with those of the now-developed countries at the end of the 19th century, there are significant differences in the urbanization process, in its antecedents and consequences. The proportion of the nonagricultural labor force engaged in manufacturing, for example, is significantly lower in today's developing countries than in their historical counterparts—approximately 40 percent in 1981 (World Bank 1983b) compared with 55 percent in 1900 (Squire 1981). One reason cited for the lower level of industrial employment in developing countries is their labor-saving bias, despite abundant labor. Moreover, in the now-developed countries, urbanization was initially the product of increases in agricultural productivity that provided capital accumulation and also created a rural labor surplus. Capital inputs were therefore available for urban development, including capital goods, which made possible increased urban labor productivity and expanded industrialization. The higher incomes that resulted served to pull surplus agricultural labor to urban areas, where the growing manufacturing sector provided job opportunities. The increase in the size of the urban population and work force led to a greater division of labor, increased specialization, easier application of technology, economies of scale, and mass production. The significant result of these developments was increased productivity, higher wages, and higher standards of living in urban areas, which encouraged rural-urban migration. Thus, urbanization in the experience of the now-developed countries was both a cause and a consequence of higher living standards.

In contrast, urbanization in the developing countries has largely taken place as a result of the push of rural inhabitants into urban areas. After World War II, mortality declined at unprecedented rates in the developing countries. This decline was largely through modern public health measures and imported medicines such as antibiotics. Death rates plummeted while fertility rates remained high. The resulting increase in natural population growth had a double-barreled effect, increasing the growth of cities and urbanization rates. The more rapid natural increase in rural areas led to population and labor force growth that could not be absorbed in the agricultural sector. Population pressure in many rural parts of the Third World and a much-reduced scope for international migration contributed to the acceleration of rural-urban migration. Urban growth was not a response to increased productivity and higher standards of living; rather, it has aggravated problems of labor absorption and its effective use, contributing to growth of slums and to the urban squalor so visible in most large cities.

Determinants of Urban Growth and Spatial Concentration

Because of their different relationship to the development process, the increased growth of cities and increases in urbanization must be carefully distinguished. Whereas *city growth rates* represent the percentage change in the absolute number of people living in a given city or group of cities, *increases in urbanization* refer to a growing proportion of the national population living in urban areas. It is also useful to differentiate between urbanization patterns that exhibit a high degree of primacy—with a large proportion of all urban residents living in the largest city (e.g., in Mexico City and Bangkok)—and more diffused patterns (e.g., in India and Malaysia). In general, urban population growth has three major sources: net migration, natural increase, and reclassification. The first two contribute the most, particularly natural increase; their relative contributions to urban growth vary in different parts of the Third World. At an early stage of development, when urbanization levels are low and urban and rural natural increase is moderately high, net migration generally contributes more to urban population growth than does natural increase. At an intermediate stage of urbanization, natural increase predominates. At a late stage, with high urbanization and low natural increase, the relationship is more likely to reverse again, to favor net migration. A large number of developing countries, particularly in Latin America, are now in the intermediate stage. A UN study (UNDIESA 1985) showed that between 1960 and 1970 in 26 large Third World cities, 37 percent of population growth was from migration and 63 percent from natural increase.

Although migration is not the major source of urban growth in many developing countries, the relatively young age of rural migrants to cities means a greater contribution to natural population increase through more births and fewer deaths. This effect tends to offset the decline in fertility rates typically associated with urban residence (Stolnitz 1984), so that urban rates of natural population increase (the difference between birth and death rates) often approximate national population growth rates. The age selectivity of the migration process and the relatively higher fertility among migrants than among urban natives also lead to young age distributions in urban areas. In most cities in developed countries, the 0-19 age group constitutes less than 30 percent of the total city population. For many cities in developing countries, the figure is over 40 percent, with enormous implications in urban areas, particularly for providing social services such as education.

Rapid population growth was once considered a major cause of rural-urban migration, with rural poverty caused by excess labor supply providing a "push" to the cities (e.g., Lewis 1954). But rural out-migration now seems less strongly associated with rural population increase than with overall economic growth, changes in agricultural productivity, and land tenure systems that promote marked inequalities in landholdings and landlessness (Oberai 1987).

Migrants move to urban areas mainly in response to better employment and income opportunities. The true determinants of urbanization and spatial concentration in developing countries are therefore to be found in the forces that determine the location of employment opportunities: the nature and pattern of industrialization, the pace of agricultural development, and the growth of transportation and communications networks.

The pace and pattern of industrial development are the most important of these, giving rise to a debate on a basic national settlement and development strategy issue: Should the emphasis be on moving industrial investment to regions where current populations are located, or should population movements become the principal means of adjustment with labor expected to move where industrial investment can locate most easily? Industries tend to locate in urban areas, especially in the larger cities, where they can benefit from ready access to capital, labor, and specialties such as financial, legal, and technical support services. Cities offer markets for industrial products and provide convenient access to other domestic and international markets through established transportation systems. The spatial concentration of economic activity and the emergence of large cities is therefore a necessary adjunct of a development process that relies predominantly on the growth of modern industry rather than on agriculture. But public policies commonly bias this basic spatial development pattern toward more rapid urbanization and more extreme spatial concentration.

Foreign exchange policies, tariffs, and industrial incentives often encourage activities of the type located in the major urban centers rather than those located in economically less progressive regions, as has happened in Brazil and Nigeria. Governmental regulation of transport tariffs and energy prices often favors large cities, as do public investment and subsidies for other urban services that influence industrial locations.

The pace of rural development also influences urbanization. The ability of the agricultural sector to absorb a growing rural labor force depends on such factors as climate; availability and distribution of land; choice of agricultural technology; demand for agricultural products; and availability of credit, fertilizers, and technical assistance. Climate and availability of land are usually immutable constraints. The Sahel region of Africa, where recurrent droughts in recent years have spurred migration and urbanization, provides an extreme example of the effect of climate. In some developing countries, particularly in Latin America and Africa, new land can still be brought into agricultural use, but in most countries there is little scope for increased agricultural employment and earnings based on newly cultivated land. The other factors impeding the expansion of agricultural employment can more readily be influenced by policy. Highly unequal land ownership distribution, especially in Latin America; slow growth and premature agricultural mechanization; and market barriers in industrialized nations have made it difficult for the agricultural sector to absorb the growing rural labor force in most developing countries, which has increased rural-urban migration.

In view of large rural populations, particularly in Asia and Africa, and the structural factors discussed above, there is still a very large potential for rural-urban migration. As noted earlier, the rural poor migrate to cities primarily because they offer more employment opportunities and higher incomes. It has proved difficult to raise agricultural output growth sufficiently to reduce urban-rural income differentials, and this situation is unlikely to change because extension of the cultivated area, a major contributor to past growth in agricultural output, is becoming increasingly costly in many developing countries or involves progressively less fertile soils. Moreover, if agricultural growth depends increasingly on intensification, the demand for urban-based inputs is likely to increase disproportionately. Rising rural incomes will also increase the demand for goods and services primarily produced in urban areas. This will not only affect the structure of urban output, it will also stimulate urban expenditures and incomes overall. Raising average rural per capita income relative to urban income levels is thus crucially dependent on lowering rural population growth, which is itself a function of natural population growth and rural-urban migration. Overall, population growth rates are now moderating in many developing regions, except sub-Saharan Africa and some Middle East countries; migration to urban areas speeds the resulting reduction in rural population growth. In light of this situation, some observers argue that what is often considered "excessive" migration, in that it multiplies problems of urban management, becomes a necessary and desirable contributor to raising relative rural per capita incomes.

Policies that protect domestic industries from foreign competition and give more favorable incentives to industry than to agriculture, agricultural credit biased toward machinery rather than labor, and neglect of rural extension and training services all tend to hamper rural development and employment—thereby pushing the rural population into urban areas—and to favor the growth of large cities over that of small towns. During the 1950s, many economists debated the correct allocation of investment between agriculture and industry. Some favored agricultural development, arguing that the agricultural sector must be made more productive to provide the necessary surplus for industrial and urban development. Without such a surplus, inadequate funds would be available for industrial and urban growth and a market for urban products and services would not develop. Counterposed to this argument was the idea that industrial and urban growth are prerequisites for a more modern and productive agricultural sector. Surplus rural labor needs to be absorbed into more productive urban activities, thereby permitting the introduction of more modern and capital-intensive agricultural practices and encouraging the diffusion of modern ideas and institutions into the traditional rural areas.

Throughout the world, in academic, and more important, in major political circles, the debate continues. Advocates of the rural and urban sectors remain convinced that most of the failures of the economy can be attributed to excessive allocation of resources to the other sector. (For a detailed discussion of this debate, see Gilbert and Gugler 1982.) Certainly urban bias is common, if not universal, in developing countries. Governments have often supported industries through special tax concessions, subsidized interest rates, tariffs, and other protective measures even as the rural sector is neglected or burdened by these very policies. In major urban areas, direct subsidies for food, plus the subsidized provision of infrastructure, health, and education services, artificially lower living costs. The price of electricity in Mexico City, for example, is about the same as it is at the source, some 1,000 kilometers away. By increasing real rural-urban wage differentials beyond what results from differences in sectoral labor productivity, these subsidies increase migration to cities (Squire 1981).

Trade policies, including exchange rate distortions, also influence the spatial distribution of population. Large developing countries with significant primary resources have typically followed the import substitution strategy, using overvalued exchange rates and taxes on primary exports to subsidize imported capital goods, with quotas or protective tariffs on imported goods that are also manufactured domestically. The effect of such policies is to turn the terms of trade against agriculture; reduce rural output; and increase rural-urban wage differentials, urban migration, and the pace of urbanization (Squire 1981).

A *minimum wage law* can also have a differentiated spatial impact that can be reinforced if coupled with a system of transfer payments to support the unemployed. Such a law provides incentives that stimulate the growth of large cities, because in developing countries the cost of living in the large cities is often higher than in smaller cities. Adoption of a minimum wage law that enforces a uniform wage rate for all cities imposes a higher real wage on industries located in smaller cities than on those located in large cities. If the demand for labor is elastic, layoffs in the smaller cities give workers an incentive to migrate to the large cities, where the minimum wage rate is not yet effectively influencing industry.

Transportation and communications networks are also important determinants of the spatial pattern of development, as they influence the movement of people, commodities, and information between regions. Public investment, taxation, and pricing and regulation of transport and communications systems can easily bias spatial development in favor of certain locations. Leaving domestic transport systems at a rudimentary stage of development encourages the location of industries in cities—usually the large ports or capital cities that have relatively good links to international and domestic markets. But if unaccompanied by other measures, improvements in domestic transport and communications may accentuate the concentration of economic activity in the large cities, as these measures lower the natural protection of industries located in small provincial centers and reduce the barriers to migration.

The main reason why manufacturing industry flocks to urban areas, particularly the large cities, is the economies of agglomeration. These economies are important analytically and from a policy perspective, as they highlight the difficulties developing countries face with any programs to encourage a more equitable spatial distribution of population. To the extent that infrastructure, capital, developed land, and management skills are not easily dispersed, only a few urban centers will be the focus of expansion. Particularly important are the constraints imposed by availability of limited infrastructure or social overhead capital. Communications, transport, universities, and harbor and port facilities are all extremely expensive; given the level of per capita income in developing countries, they are difficult to duplicate outside the large cities, at least initially. Business decisions to invest are also influenced by the availability of specialized business services such as brokers, banks, trade associations, consulting services, laboratories, and equipment leasing. Most of these facilities, so essential for modern industry and business, require a minimum population size for their profitable operation.

Spatial Concentration, Economic Efficiency, and Growth

As observed above, urban growth often gives rise to economies of scale. Industries benefit from the concentration of suppliers and consumers, which allows savings in communications and transportation costs. Large cities also provide big, differentiated labor markets, which often help to accelerate the pace of technological innovation. Large cities enable economies of scale for such services as water supply and electric power to be exploited. Set against these benefits, unemployment is generally higher in urban than rural areas. Air pollution, congestion, social disturbances, crime, and similar problems are also thought to increase disproportionately with the growth in city size.

Most studies show that at the lower end of the size range, economies of scale increase rapidly as a city expands, but beyond a certain size the additional gains diminish rapidly (see Renaud 1981). But none of the studies has yet been able to pinpoint at what city size the losses created by congestion and environmental deterioration equal or exceed the benefits of agglomeration, so we cannot say, on these grounds, that 3 or 6 or 10 million people in a metropolitan area are too many.

The judgment that major cities have become too large normally rests on the assumption that urban diseconomies—say, pollution or traffic congestion—have become so severe that the only answer is to deconcentrate both population and economic activity. Such an argument has several failings. First, it does not recognize that diseconomies are only one side of the argument; if urban agglomerations generate still greater urban economies, then the balance of economic advantage rests with spatial concentration. Second, even if diseconomies do exceed economies, the best policy may be to improve urban management rather than to deconcentrate population and employment. To reduce traffic congestion, for example, the best policy may be to improve public transport, cut the use of private cars, or introduce parking meters. Air pollution can be reduced by the physical removal of polluting industries, but it can also be cut by fining errant companies. As for unemployment, governments are sometimes responsible for reducing the absorptive capacity of cities by intervening in labor markets (e.g., through minimum wage legislation or licensing requirements and restrictions on small businesses) and by pursuing inappropriate pricing policies for public services. National economic policies—for example, providing fiscal incentives and low-interest loans to promote capital-intensive industry—may also exacerbate urban problems by encouraging rural-urban migration without creating enough new urban jobs.

In the interests of equity, it may be argued that industrial activities should be dispersed to smaller towns and rural areas throughout a developing country, but the efficiency costs of such a policy may well be prohibitive. Many economists believe that government intervention in the distribution of economic activity is likely to waste scarce capital resources and thereby slow the rate of economic growth. In the longer term, the country will thus be less able to redistribute income and remedy the problems of poverty. Regional balance and urban deconcentration should therefore be left until a country has achieved a higher level of development. To achieve a high rate of economic growth, further concentration of population into a few large metropolitan areas cannot be avoided.

This view is supported by the finding that large cities are often more efficient and innovative than other urban centers. Several studies indicate that industrial productivity is relatively high in large cities, even with allowance for differences in capital per worker and enterprise size (Kasarda and Crenshaw 1991; Richardson 1973; Rocca 1970). Evidence also suggests that per capita costs of social overhead capital tend to fall with increasing city size or at least do not rise (Richardson 1973). These findings thus support the view that urban economies exceed urban diseconomies, even in today's large cities, and no prima facie case favors deconcentration.

Such a view remains controversial, however, and the evidence on which it is based has been criticized on many grounds. The efficiency of large cities has also been questioned. Gilbert (1976) argues that evidence of higher productivity in large cities should not be attributed to agglomeration economies alone, for such "economies" may derive from better urban infrastructure or higher quality labor. In the latter case, it may be argued that higher productivity in large cities is to some degree achieved at the cost of lower productivity in smaller cities; if equivalent infrastructure or labor were available in small and medium-sized cities, then their productivity might well rise. In addition, productivity among private firms in large cities may appear high because private companies are often subsidized indirectly by the state. If firms had to bear the full externality costs they impose, the higher productivity of large cities might well be less pronounced. Moreover, if firms were forced to pay some share of the diseconomies they create, they might find large cities less attractive and many might decide to move to the intermediate cities. This process would reduce the apparent differential in industrial productivity between small and large cities.

Even if urban economies do outweigh the diseconomies in large cities in developing countries, are the overall benefits equitably distributed? One can argue that urban diseconomies most affect the lower-income groups, who are least able to escape them. The middle- and upper-income groups often have the resources and knowledge to move to different areas, command better public services, and influence political decisions. Industrial zones are designated to keep pollution and trucks out of high-income areas. Urban renewal schemes rarely displace the rich but often dislocate low-income communities. Public roads and telephone, water, and electric services are often superior in high-income areas, and where the public sector cannot provide adequate services, as in health and education, the wealthy can resort to the private sector. The operation of the land and housing markets guarantees that the wealthy gain most from uncontrolled speculation.

Those supporting decentralization thus argue that in the short run we may achieve the objectives of growth and efficiency by encouraging population concentration in a few urban centers, but in the long run, the results are more likely to be mass poverty, greater individual and regional income inequalities, and social unrest. Thus the state must redistribute income and wealth among regions and income classes, not only because extreme income inequality is morally wrong, but because it is impossible to achieve national unity and prevent social unrest in the face of glaring regional and family income disparities. They often suggest three kinds of policies:

- Those that aim at transforming the rural economy, thereby slowing the rate of urban expansion
- Those that aim at limiting the growth of large cities through the control of migration
- Those that try to slow the growth of large cities by stimulating the growth of intermediate cities or establishing new urban centers

Any policy that transforms the rural economy affects the nature and pace of urban development. Policies such as redistributing land to the poor may reduce urban-rural income disparities and slow urban growth by raising agricultural incomes. Other rural programs may have the opposite effect: The green revolution and other efforts to spur productivity via incentives to commercial agriculture have accentuated landlessness and probably stimulated, rather than reduced, the flow of city-bound migrants (Griffin and Ghose 1979).

Many Third World city administrations have felt the need for direct measures to reverse the flow of migration and stop or discourage urban migration. Such measures have usually included administrative and legal controls, police registration, and direct "rustication" programs to move urban inhabitants to the countryside. But only a few countries, such as China and Cuba, have been able to reduce the flow of migration to large cities through residence permits and legal restrictions. Elsewhere, such attempts have been largely unsuccessful because legal restrictions are difficult to enforce, licenses are easily forged, monitoring entire cities is costly, and evicted persons can return (Oberai 1987).

By far, the most common response to urban and regional imbalances has been a program to stimulate the economies of small and medium-sized cities. The prime candidate for inclusion in such a program is often the industrial sector. Existing or new companies are persuaded to locate their plants in peripheral regions through a combination of tax incentives, provision of infrastructure, construction of industrial estates, and occasionally, compulsion. In general, the stick has been used much less than the carrot; most governments have feared that too strong a program of deconcentration would dissuade foreign firms from investing in the country, lower efficiency in the industrial sector, or offend important national business interests.

Various catalysts have lured industries to new locations. In Brazil, major tax incentives attracted numerous companies to the principal cities of the poverty-stricken northeast. Experience shows, however, that unless the incentives to do otherwise are extremely attractive, manufacturing companies have a tendency to locate in the areas closest to their preferred locations. Thus, in northeast Brazil, new industry was located only in the three largest cities, and in Peru, new industrial estates attracted companies only to the most prosperous and attractive city, Arequipa (Gilbert 1974). India's Second and Third Development Plans led to the establishment of 486 industrial estates across the country, but only those close to the large cities prospered.

In some countries, industry has been encouraged to move from the large cities. The issue here is whether such deconcentration has had a positive effect on the poorer regions. Often industrial dispersal has increased regional output, but created few jobs. Indeed, deconcentration programs have tended to suffer from the same weakness as national industrialization programs: Capital-intensive technology has not created many local jobs or much demand for local inputs in the way suggested by growth center theory. As a result of weak "spread" effects or substantial "backwash" effects, the regions beyond the immediate vicinity of the growth centers receive little in the way of positive economic or social benefits. Large-scale industrialization within a growth center is an ineffective means of developing a poor region in the absence of fundamental changes in the agricultural economy, the marketing system, and landholding patterns. Industrial deconcentration may reduce pressure on the metropolitan areas, but unless accompanied by other, often more radical, programs, it brings little benefit to the poorer regions.

The problem is that although the poor regions may benefit from regional policies, the poor inhabitants of those regions do not benefit sufficiently. Too often, the benefits go to the industrial and large-scale agricultural groups in the region. A reduction in interregional disparities is then accompanied by a growth in intraregional disparities and personal inequalities.

Although major regional programs—and indeed programs for planning the major cities—may be essential, they are not in themselves sufficient. The type of development strategy that is followed is critical. When urban and regional policy goals have conflicted with the paradigm of national economic growth, the latter has usually been the winner. Many governments have established regional programs espousing goals of greater equity and regional balance. Almost invariably these policies have been counteracted by national programs accentuating regional disparities and encouraging the growth of major urban areas. Of course, introduction of regional programs may have reduced the level of spatial concentration that would have resulted from the national programs alone. But in general the weakness of regional and decentralization programs suggests they are never given priority or intended to achieve a real balance.

The case of Lagos demonstrates well how macropolicies can submerge decentralization efforts. Over 90 percent of total net subsidies granted to Nigerian industries benefit ones located in Lagos, which could not be expected to have a positive impact on industrial decentralization. Decentralization policies have clashed with other policies in India, too. The Indian government instituted a policy licensing no new industrial activities in its five large urban centers. Simultaneously, export promotion was a declared policy. Three large urban centers, the ports of Calcutta, Bombay, and Madras, are prime locations for export-oriented industries. The effect of the licensing ban on new urban industry is unclear, but it cannot have been favorable. A possible side effect may have been driving activities underground, which would have attenuated the unfavorable effects on exports but also frustrated the spatial objective of restraining major urban center growth.

Although major differences of opinion exist as to the desirability of spatial deconcentration and the possibility of good metropolitan planning, most writers agree on certain points. First, the concept of an optimal size of city is not useful. Because city organization is at least as critical as size, and the nature of metropolitan economies varies considerably across the globe, no single city size can be optimal. If a case is to be made in favor of deconcentration, it must be made on the basis that certain goals, such as equity and national unity, can be better achieved in a more deconcentrated settlement system. Second, whether or not deconcentration is favored, more effective and fairer urban planning is a necessity in large cities because, as seems clear even in China or Cuba, metropolitan growth cannot be stemmed for long. Such planning requires the implementation of policies and programs to improve conditions of the urban poor. Third, more should be done to compensate those who suffer from noise, pollution, and congestion. If large cities do indeed generate vital urban economies, companies will remain, even if they are forced to pay the costs they impose on other city dwellers. Finally, rather than directly intervening in population distribution policies, governments should try to correct the biases of national development policies that have implications not only for industrial location patterns but also for the regional distribution of purchasing power. Industrialization policies should be consistent with overall macroeconomic policies and concentrate on improving peoples' welfare regardless of where they live.

Concluding Remarks

Rapid urbanization and concentration of economic activity in a few locations, particularly large cities, are inevitable outcomes of economic and industrial development. Without information on the relative costs and benefits of urban versus rural development and of the growth of large versus small cities, it is difficult to determine any preferred rate of urbanization and the best spatial distribution of economic activity in any given country. It is also impossible to judge the optimal size of a city. In any case, economic efficiency is not the only aspect of urbanization with which governments are concerned. In most developing countries, particularly those with distinct regional or ethnic interests, it is important to maintain a degree of balance between regions and between rural and urban development; hence, some attempt to slow the urbanization process and spread economic development more evenly across regions may be politically necessary, even if its economic desirability is less obvious.

In terms of slowing urban growth, the range of policy instruments for controlling migration is quite limited. Instead of directly intervening in population distribution policies, governments should try to correct the biases of national development policies that have implications for industrial location patterns and regional income inequalities. Yet, removing urban bias in national policies and regional development efforts is unlikely to have much impact on facilitating population deconcentration from large cities, at least in the short run. Therefore policies to improve internal efficiency in large cities must be vigorously pursued as the most reasonable short-term strategy.

4

Urban Land and Housing Issues Facing the Third World

ELLEN M. BRENNAN

There are complex linkages among population and land and housing issues. Throughout the developing world, rapid population growth from continuing high rates of natural increase and rapid in-migration has resulted in a high rate of consumption of urban land. Many of the world's cities will nearly double in size over the next several decades. Boundaries constantly spread outward, resulting in what in Mexico City is vividly described as the "urban stain."

Most Third World cities have burgeoning peripheries that have grown almost overnight by means of invasions or are the result of gradual accretion. Informal urban periphery development frequently has created sprawling, low-density settlements that are difficult to service and have poor sanitary conditions. Except for a few authoritarian regimes, very few governments have succeeded in controlling this unplanned urban growth. Past community upgrading efforts have usually concentrated on physical components, ignoring the existence of marginal settlements that persist mainly as a symptom of deeper economic and social factors directly traceable to large-scale poverty (Gilbert 1990).

The current situation in land and housing markets has been a drag on development in many Third World countries. The fundamental issue in most cities is more complex than a limited supply of urban land. The issue is more one of speculation and inefficient land use, with large vacant areas in the central city held by speculators, forcing workers to live on the periphery and commute long distances to the central city.

Construction of high-quality, heavily subsidized, high-cost housing units has been a failure. Projects in many countries were devised without considering needs of the target groups, resulting in housing too costly for the intended beneficiaries.

Land Development Issues

With some Third World cities growing at 5-10 percent per year—and at rates as high as 20 percent or more on the urban fringes—land in urban areas is being consumed at a rapid rate. In Cairo, for example, an estimated 1,200 hectares of agricultural land are lost annually to urban encroachment. Mexico City has been losing at least 1,000 hectares of agricultural land and 700 hectares of forest land each year (UNDIESA 1990a, 1990b).

The extraordinarily adaptive nature of urban land markets under pressures unprecedented in the developed countries is surprising. Doebele (1987) explained the adaptability of urban land markets by the "porous" qualities of most developing country cities at the end of World War II. Most cities grew in an unplanned, loosely structured fashion that left many vacant interstices. Central city dwellings were often located on large plots suitable for construction of additional structures. Public authorities frequently owned large tracts in central areas that could be used for the construction of shelter. As urban growth rates accelerated, the price of central city sites increased dramatically, becoming far too expensive for low-income residential occupancy and creating pressures to expel the existing residents. On the periphery, entrepreneurs could often purchase farmland to subdivide and resell. Despite rudimentary services, such illegal subdivisions offered enough of a sense of security of title to encourage their owners to use them as economic resources (Doebele 1987).

This situation is now rapidly changing. Many options previously available to low-income urban people, such as unused public land, are disappearing rapidly even as access to peripheral land is becoming increasingly restricted. Indeed, vacant land on the urban fringes and elsewhere is being assembled and developed by corporate developers, legally and illegally. As developers and speculators tie up large areas of land on the urban fringes, land markets are becoming bottlenecks in the development process (UNCHS 1989b). Moreover, governments are frequently required to service such developments at the expense of efficiently planned infrastructure elsewhere.

Even as demand for land grows, the supply in most developing country cities is genuinely and artificially limited. The problem of land supply is generally independent of the type of ownership. It is as severe in cities such as Karachi and Delhi where much of the land is publicly owned, as in cities like Bangkok, Metropolitan Manila, or Seoul where most of the land is privately owned (Brennan and Richardson 1989). In all cases, land prices have risen much faster than the consumer price index, exacerbating the difficulty of acquiring land for low-income housing. In some cities, this price increase is partially explained by special circumstances, such as the absolute scarcity of land not subject to flooding (Dhaka), mountainous terrain (Rio de Janeiro), purchases by nationals returning from working in the Middle East (Karachi and Dhaka), or the preservation of a large greenbelt (Seoul).

A more general consideration has been the appeal of land as an investment. Again, speculation seems as rampant in cities where much of the land is publicly owned as in cities where private developers dominate. In Karachi, where 80 percent of the land is publicly owned, there has been widespread speculation, mainly because of the large number of serviced sites developed by the Karachi Development Authority and sold well below market prices (UNDIESA 1988a). In Bangkok, however, where 80 percent of the land is privately owned, aerial photography confirms some 1,650 square kilometers of vacant land in and around the Bangkok Metropolitan Area (UNDIESA 1987a).

Because of expected demand for land in most developing countries and a lack of alternative investments, land is often seen as the only reliable avenue for investment. Where remittances from overseas earnings have been used for land purchases, this tendency is even more pronounced. Dhaka, the world's poorest mega-city, has seen intensive urban land speculation. An estimated one-third of expatriate remittances have gone for land purchases. Land prices have risen about 40-60 percent faster than prices of other goods and services and are now completely out of line with income levels (UNDIESA 1987b).

In Cairo, remittances from abroad have also figured in most residential land transactions, fueling the rise in land costs. An early 1980s study concluded that unless Cairo authorities either created attractive alternative investment opportunities, effectively taxed potential capital gains on land sales, or effectively prohibited private land transactions by massive land banking or expropriating developable urban land, the investment demand for land would likely continue at a high level, inflating land prices beyond the means of most Cairo residents (UNDIESA 1990a).

In addition to creating an artificial scarcity and escalating prices, speculation has other undesirable effects. In many developing country cities, large betterment values—often from improvements in accessibility through public-sector investments—have accrued to private developers, even as the social costs of speculation (e.g., leapfrog development) are passed on to the public. In other cities, speculation has distorted the residential land market and caused a highly dispersed, discontinuous pattern of urban development.

Devising equitable and efficient land development policies is a major challenge facing policymakers in Third World cities. Market mechanisms alone are unlikely to create an efficient allocation of land uses in cities, so interventions (physical, legislative, administrative) are needed to increase efficiency, distribute benefits more equitably across income groups, and reduce congestion, pollution, and other negative effects of inappropriate land development (UNCHS 1989b). Even those societies most oriented to free enterprise have increasingly imposed measures of public control over urban land use.

The degree of intervention varies by country. Cities like Bangkok, Dhaka, and Metro Manila have had virtually no effective measures to influence or control land development; Seoul represents the opposite extreme. Independent of the degree of intervention, there are similarities. Many cities have master plans prescribing directions of urban growth, but these plans rarely are realized and languish in metropolitan planning offices as irrelevant documents (Brennan and Richardson 1989). The reasons are simple. Population projections underpinning the master plans are often far off the mark, and land uses soon diverge widely from the plan. Most master plans are formulated in an inflexible manner, not allowing for readjustments for changing conditions.

Land use controls have been largely ineffective. They often reflect engineering or architectural concern with order and the clear delimitation of land uses inherent in developed country town planning ideals (UNCHS 1989b). Desirable standards of plot size, infrastructure, and so forth are typically set at levels higher than the majority of low-income residents can afford. Moreover, there is little or no recognition of the limited enforcement powers of planning departments in most developing country cities.

In a review of urban land policies, Doebele (1987) argued that the key problems of urban land markets are increasing the total supply of land accessible to low-income households and making better use of the substantial underused land areas in many large cities. He listed three major approaches: direct public actions to increase land supply, joint public-private actions to increase land supply, and more efficient use of existing land resources.

One major direct public action to increase urban land supply is reducing its attractiveness as a vehicle for capital storage. Several governments have tried this approach, with limited success. In India and Pakistan, where government has long tried to control speculation, restrictions on multiple holdings, size of holdings, and the warehousing of vacant land have been circumvented by tactics ranging from litigation to substituting family proxies.

India's 1976 Urban Land Ceiling and Regulation Act imposed a ceiling on vacant land holdings for individuals and companies. In the large cities such as Bombay, the ceiling was only 500 square meters. Landowners were required to register their holdings and surrender excess vacant land to the government for compensation fixed at a standard rate (UNDIESA 1986a). In all of India, only 1.5 percent of the land declared as surplus has been acquired to date under the act (UNCHS 1989b). Not only has this limited the supply of land that can be offered by the public sector, but large amounts of privately owned "surplus" land have been frozen in litigation or administrative proceedings for most of the period since enactment of the act.

Karachi's government has attempted to control speculation by limiting the number of plots an individual can own. The law has been easily circumvented by use of family proxies. Moreover, Karachi's property and capital gains taxes have aided investors in holding plots they never intended to occupy. For example, some 80,000-100,000 of the 260,000 plots developed by the Karachi Development Authority during the 1970s were held for investment and lay vacant 10 years later (UNDIESA 1988a).

Legalizing land tenure is another way to increase the urban land supply. Lack of tenure security is a major bottleneck in the efficient functioning of land markets. An estimated 20-40 percent of urban households in developing countries live on land to which neither they nor their landlords have title (Mayo, Malpezzi, and Gross 1986). In many cities the figure is considerably higher.

Tenure security can have powerful effects on dwellers' incentive to invest in housing and associated infrastructure; in many developing countries it is a precondition for access to formal mortgage financing. The degree of property rights needed to achieve a perception of security varies with circumstances and culture. Different strategies to increase tenure security have diverse costs and benefits. Although it costs a government little to confer tacit approval on existing settlements, formal registration with cadastral mapping can be very costly. Tacit approval may be sufficient to stimulate some investment and encourage maintenance, but more formal systems are usually required to obtain funds from mainstream financial institutions (World Bank 1989a).

Many cities have met problems in legalizing tenure, particularly as to cost recovery. In Karachi, the government policy has been to regularize all squatter areas (*katchi abadis*) located on provincial or local council land and built upon before 1 January 1978 (later extended to 31 March 1985). Squatters received 99-year leases that could not be transferred for 5 years except by inheritance. In return for receiving ownership rights, a onetime charge was levied on the plot holder, plus a nominal annual rent.

Each regularization and improvement scheme in Karachi was supposed to recover a major portion of the development costs. In practice, there were problems in achieving cost recovery. Although charges for the land and its development were very low, the pace of recovery was slow. A pilot project in one township found that in over 15 months, less than 10 percent of residents came forward to obtain title. Generally, unless households wanted to sell the land after the 5-year waiting period, people were content to maintain the status quo. Subsequently, the government reduced the development fee. But rather than being encouraged to obtain title, most residents waited to see if rates would come down further or be abolished entirely (UNDIESA 1988a).

Another dilemma in granting tenure is the fear that granting any form of tenure will be tantamount to legitimizing an illegal act, encouraging further squatting. Parallel to the legalization program, Karachi authorities made aerial photographs and announced a freeze on the physical limits of *katchi abadis* that could be regularized, but the freeze may be impossible to enforce because of the political climate (UNDIESA 1988a).

In some cities, granting security of tenure has produced *upward filtration*—middle-income households buying out lower-income residents (Doebele 1987). In Mexico City, large numbers of low-income families have invaded state-owned land in inhospitable areas on the periphery. Public authorities tolerated and even legitimized this process, concentrating on programs of massive ex post regularization whereby low-income residents who invaded land or purchased lots in irregular subdivisions could pay the government and receive title. Ironically, in spite of official pronouncements seeking to control Mexico City's urban sprawl, the regularization program has fostered the spread of unplanned urbanization. Once land on the urban fringes has been regularized and infrastructure introduced, land and housing prices have soared, forcing lower-income residents to sell and move to cheaper, unserviced areas farther out on the periphery (UNDIESA 1990b).

Governments have also tried to increase the supply of urban land through modernization of cadastral and land registration systems. A major reason why local administrations in most cities have not coped successfully with growth is because they do not know what is going on in their local land markets (UNCHS 1989a). The information base in many cities is improving, particularly with the aid of aerial photography. But most cities lack accurate, current data on land conversion patterns, number of housing units (informal and formal) built during the past year, infrastructure deployment patterns, subdivision patterns, and so forth. Often, city maps are 20-30 years old and lack descriptions of entire sections, particularly the burgeoning periurban areas. Without information, planners and policymakers have operated by making assumptions or using often inappropriate standards from other nations to develop land policies. Unless assumptions or standards are on the mark, plans and policies are ineffectual at best and destructive at worst (UNCHS 1989a).

Land registration has always been a serious problem. Information on who owns what is typically poor. The legal and administrative systems to establish, record, and transfer title are usually inadequate, and squatter settlements boost uncertainty about property rights (Mayo, Malpezzi, and Gross 1986). For land markets to work properly, transactions must be registered. Once facts are recorded, land can be bought and sold with fewer obstacles; that helps the pace of development (World Bank 1989a). Better registers, besides aiding in improving property tax systems, can also bring in substantial transaction fees, as worldwide some 10 percent of urban property changes ownership each year.

Bangkok has a good example of a successful land-titling project. Property values were a major area in which Bangkok's growing fiscal capacity had not been tapped. The Bangkok Metropolitan Administration had been using cadastral maps that were 40-60 years old. Despite recent, substantial increases in land and property values in Bangkok (often from public investments such as highways), this betterment was not even partially recaptured. To remedy the problem, the government, aided by the World Bank, undertook a land-titling project aimed at producing up-to-date maps, introducing new surveying technology, strengthening the Central Valuation Authority through technical assistance, and eliminating leakages in the current valuation system

(UNDIESA 1987a). The photo maps were sufficiently detailed so that few parcels of land escaped identification for taxation. Before the project, an estimated one-third to one-half of Bangkok's 1 million land parcels were not linked to a tax record. In 1988 alone, Bangkok's land titling brought in $200 million (World Bank 1989a).

In addition to direct public actions, some governments have considerable experience with joint public-private actions to increase land supply. Land readjustment has worked well in Japan and the Republic of Korea. Seoul has a long history of relatively successful intervention in the urban land market. Seoul's land readjustment program was enacted in 1937 to deal with unplanned urban growth on hillsides and valuable agricultural land. Land readjustment involved the public assembly of small parcels of undeveloped land on the urban fringes without monetary compensation for the owners. Typically, the land was subdivided and the portion needed for streets, recreational areas, and other public uses was set aside. Authorities retained some of the land and sold it at market prices to cover infrastructure costs. Finally, a parcel of land in proportion to the landowner's share and located as close as possible to the original holding was returned to the landowner, who could then make a sizeable profit by selling the remaining improved land. The land readjustment program is an example of successful intervention in the urban land market, as it involves only short-term public-sector landholding and results in significant improvements. The program is not necessarily transferable to other developing countries, however, as it requires major administrative infrastructure, a well-designed legal framework, an effective political consensus, and speedy implementation (UNDIESA 1986a).

Another example of public-private cooperation broader in scope than land readjustment is the Guided Land Development Program (GLD) in Jakarta and other Indonesian cities. Essentially a simplified form of land readjustment, GLD tries to guide development to the fringe areas east and west of Jakarta by providing serviced urban land on a large scale affordable by low- and middle-income households. Areas are selected based on available land, potentially easy provision of main entry roads, adequate drainage conditions and water catchment, and an existing housing stock that is not too dense. Development plans and actual layouts for the areas are prepared with the participation of area residents and landholders. Decisions are made at the outset about public rights-of-way, and key parcels for essential community facilities are bought at public expense. To keep prices from rising sharply, GLD aims at providing as much or more land than demand requires. Improvements to minimum acceptable levels are staged incrementally so that land values do not escalate beyond the level of affordability of the majority of households (UNDIESA 1989a).

Housing Issues

Housing is not only an economic good but also a basic human need. The stakes in making a good housing policy are high, as the sector forms an important part of a nation's real wealth. Indeed, the value of a country's housing stock is typically larger than its gross domestic product and often represents its largest single component of reproducible wealth, comprising between 30 and 50 percent of total reproducible assets (World Bank 1988a).

Pressures of urban growth in the developing world have put enormous strain on urban housing markets. During the early 1980s, nine new households were formed for each permanent dwelling built in low-income developing countries (World Bank 1988a). The gap between supply and demand in most developing countries is widening. In Thailand, against estimated demand for 300,000 urban housing units, the National Housing Authority could build only 6,000-7,000 units annually (UNDIESA 1987a). In Madras, some 6,000 legal housing units have been produced annually by government agencies and the private sector—against estimated annual demand for 30,000-40,0000 new units (UNDIESA 1987c).

In some instances, however, housing problems are more complex than simply supply versus demand. Studies confirm the apparent anomaly of a large and growing housing surplus in Cairo during a time of widely perceived housing shortages. Indeed, chiefly because of informal-sector contributions, Cairo's housing stock has expanded not only at a rate high enough to accommodate new household formation and in-migration, but also to accommodate some moves by established households. Although Cairo's housing market has major problems, they are related more to its specific features (e.g., rent controls) and housing costs than to the inability to find shelter (UNDIESA 1990a).

Public Housing

Until the early 1970s housing policies in most developing countries typically followed the model of industrial nations, relying on heavily subsidized blocks of public housing targeted to special groups. The number of units constructed was constrained by high building standards and resultant high costs. Standards were often adopted from colonial authorities or developed countries; sometimes high standards were demanded as proof of modernization and economic progress (UNCHS 1989b). As most of the intended residents could not afford the units, rents were usually subsidized. Despite large subsidies, however, public housing often went unoccupied for long periods because of poor location, inadequate infrastructure, or cultural unacceptability (Mayo, Malpezzi, and Gross 1986).

To cite a few examples, in Jakarta, public housing units were generally constructed in peripheral areas and reserved mainly for public servants and the military. Some units were offered to the general public, but these were expensive and lacked appeal. Living vertically was an alien way of life for most Jakarta residents. Not only were there few opportunities for employment in the area, but self-employed households were prohibited from selling food or operating cottage industries in the new high-rise units (UNDIESA 1989a).

In Bangkok, the large, multistory apartment blocks constructed by the National Housing Authority (NHA) in the late 1970s were neither culturally acceptable nor financially available to their targets. The illegal transfer of apartments for "key money" became common. Because of poor construction and lack of maintenance, many public housing complexes constructed under NHA auspices are now classified by it as "potential slum areas" (UNDIESA 1987a).

Informal Housing

Because of deficiencies in formal housing, cities throughout the developing world are being built primarily by the informal sector. Cairo provides a good example. Informal housing is estimated to account for 84 percent of all housing units built in Cairo from 1970-1981 (Abt Associates n.d.). Although informal housing is similar in design and materials to formal housing (new informal housing in Cairo is of far better average quality than older housing), it generally contravenes either building codes or zoning laws. One-half to two-thirds of the informal housing units added to Cairo's housing stock from 1976-1981 were additions of floors to existing buildings. The remainder were intrusions on privately owned agricultural land, usually with irregular layouts. Utilities and infrastructure in Cairo are typically put in place after the housing is constructed, at considerable added cost (UNDIESA 1990a).

Mexico City's situation is fairly similar. More than half of its housing stock has been built and financed directly by low-income groups, mainly on land of uncertain tenure. Although tenant-built informal-sector dwellings usually involve no immediate public outlay, for other reasons these unregulated developments have been expensive. For example, the political pressure of squatter neighborhoods often forces authorities to provide basic services at greater expense than in planned sites and with little if any recovery of investment costs. Moreover, programs to regularize land tenure ex post and to introduce real estate taxation have been difficult and costly as well as often displacing the original occupants (UNDIESA 1990b).

Squatter Settlements and Slums

There is no simple system of classifying Third World low-income, urban residential areas. Squatter settlements may be defined as illegal land occupation. Not only is legality difficult to define, but many unplanned settlements mix legal and illegal characteristics (UNCHS 1982). Squatter settlements are largely built by the inhabitants by whatever means come to hand and usually lack public utilities and community services. No single term describes the various types of settlements, often known by local terms such as *barriada* (Peru), *favela* (Brazil), *katchi abadi* (Pakistan), *bidonville* (French-speaking Africa), shanty town (English-speaking Africa), and so forth. Such settlements are the most conspicuous sign of how Third World land markets work, but they do not define what is wrong with those markets (Mayo, Malpezzi, and Gross 1986).

Slums are areas of older housing that are deteriorating because of overcrowding and lack of services and maintenance. Older, established slums are typically in city centers in large, well-defined areas, such as India's Old Delhi, and in neighborhoods built for low-income workers during a city's industrial expansion (Calcutta's *bustees*, Bombay's *chawls*, the *vecinidades* of Mexico City). With age, some squatter settlements assume traits of slum areas.

The prototypical squatter settlement, built by owner-occupiers who invaded the land illegally, is found in Latin America. In Asia, squatter settlements usually have grown by gradual accretion rather than planned invasion (UNCHS 1982). For both, land occupation is illegal, but invasions are more organized and occur very rapidly, often within a few days.

Mexico City invasions come so rapidly that squatters are known as *paracaidistas* (parachutists).

Squatter settlements typically grow either through expansion or increases in density in a limited area. Slums usually cannot expand because the surrounding urban land is already occupied, so they grow mainly through the addition of shanties on vacant plots and rooftops, or by converting buildings to rooming houses. Squatter settlements are less likely to develop in cities with ample low-income tenement housing. In São Paulo, with much dilapidated older housing (*corticos*) in central areas, *favelas* are a fairly new entity.

Although squatters' ingenuity and imagination in solving their own shelter problems under unfavorable conditions have frequently been praised, the vast majority of their dwellings are unfit for human habitation. Most squatter housing has an estimated life span of under 5 years. Over 80 percent of dwellings in selected squatter settlements in Lima, 90 percent in Seoul, and 100 percent in Kuala Lumpur had very limited durability (UNCHS 1982).

Third World squatter settlements exhibit several types of construction. Some informal housing even in squatter areas is of surprisingly high quality, erected with durable materials like concrete blocks and roofed with corrugated iron, asbestos, zinc, or aluminum. Another type is based on traditional rural construction materials—mud and wattle, mud bricks, or compact earth blocks (UNCHS 1982). Finally, much squatter construction in developing country cities consists of buildings made from a wide variety of scrap, scavenged, and recycled materials (scavenged timber, tin scrap, cardboard, or plastic sheets).

Governments sometimes ignore slum and squatter settlements and allocate public resources elsewhere. Similarly, governments often exclude squatter settlements from public utilities and health and social services. Bulldozing squatter settlements was common during the 1960s and early 1970s. Land policy often went no further than the rule that if squatter settlements were growing, evict the squatters (Mayo, Malpezzi, and Gross 1986). Such policies were rarely effective. They displaced rather than eradicated the squatter settlements. Moreover, they failed to consider that squatter housing was a large part of the poor's capital stock.

Outright destruction of squatter areas is now relatively uncommon, except in cases where squatters live in areas unsuitable for upgrading. Karachi provides an example. Although the government policy is to regularize and gradually improve *katchi abadis,* the law stipulates that

scattered clusters of under 40 households and encroachments on land designated for future public amenities (e.g., mosques) or in dangerous areas (low-lying flood-prone areas) cannot be regularized and must be removed (UNDIESA 1988a).

Sites-and-Services Schemes

The earliest sites-and-services schemes (e.g., Chile, Peru, Kenya, South Africa) were undertaken in the 1940s and 1950s, largely without external assistance. In the early 1970s international support began in earnest. The World Bank (1989b) began sponsoring the schemes in 1972. These early sites-and-services projects consisted essentially of a "basic needs" approach that emphasized lowering standards to the bare minimum by providing shelter that could be upgraded over time. Typical projects included land tenure, selected trunk infrastructure to connect the areas with existing utility and road networks, on-site infrastructure, core houses (ranging from a simple wall with utility hookups to completed buildings), social facilities (e.g., health clinics), and funding for the plots, core houses, and building materials.

Some sites-and-services projects have been relatively successful but failed to achieve their broader objectives. In evaluating Bangkok's three major ongoing sites-and-services projects, Thailand's National Housing Authority (NHA) concluded they had not fulfilled all of their original objectives. One project, intended as the basis for a new town, was frustrated by a lack of jobs nearby. A second sites-and-services project included plans for an industrial estate, but most residents continued to commute to Bangkok. The third project was to have developed into an autonomous new town, but NHA acknowledged that by the time it was completed it would be engulfed by the Bangkok Metropolitan Area (UNDIESA 1987a).

Area Upgrading

Sites-and-services projects represent a considerable improvement over previous shelter strategies, but they are based on new construction, and even under the most optimistic scenario, they cannot hope to reach the majority of low-income urban residents. Area-upgrading projects, however, build on the existing informal housing stock in which the poor already live and thus are in a better position to target shelter production directly to them. Typically, area upgrading aims at improving infrastructure in a comprehensive package with water, sanitation, drainage, solid waste removal, roads, and footpaths. Some form of tenure security usually has been part of slum improvement projects.

Though one on the world's poorest cities, Calcutta has mounted a large area-upgrading program and has made much progress in addressing some of its most serious infrastructure deficits through alliance of the strong Calcutta Metropolitan Development Authority (CMDA) and the International Development Association. Under the aegis of the CMDA, major area upgrading involving paved internal roads, electrification, and water standpipes and sanitary latrines have improved living conditions for nearly 2 million persons (UNDIESA 1986b).

Jakarta's Kampung Improvement Programme (KIP) is widely acknowledged to be a successful area-upgrading program. KIP's basic principles are to implement improvements that are simple to make, even if marginal, in the living standards of as many residents as possible. KIP funds improve public facilities—upgrading roads, canals, and water supplies and building social welfare facilities and communal bathing and washing facilities, rather than private accommodations. Although informal, unplanned, and until recently, unserviced, the majority of *kampungs* have some form of legitimate tenure. During the program's first 5 years, KIP was funded from the city's resources, with standards set low to stretch funds as far as possible. In 1974 the World Bank extended a loan to expand the program, with a second loan approved in 1977 and a third in 1979. By the early 1980s, KIP had improved more than 500 kampungs and had provided basic services for an estimated 3.8 million residents (UNDIESA 1989a).

Because of KIP's longevity, it has been possible to conduct several longitudinal impact studies. KIP triggered substantial private investment in home improvement and led to significant property value increases. The National Urban Development Study found home values increased 2.6 times the amount of money invested in home improvement (UNDIESA 1989a).

Slum Upgrading

Scant attention has gone to renovating slum housing in cities' deteriorating central areas. The central location of slum areas poses problems, particularly in market economies where land prices increase toward city centers, making these sites potentially valuable for redevelopment. Commercialization and gentrification are major threats to the inner-city poor, often resulting in their expulsion to periurban squatter settlements. Governments rarely intervene to counteract these trends. In fact, an analysis of legislative and regulatory measures shows that public-sector initiatives generally intensify the phenomenon (UNCHS 1984b).

Redevelopment activities in inner-city areas, either through public- or private-sector intervention, have been unsuccessful in providing alternative, improved housing for the poor. The interesting lesson is that this phenomenon is pervasive in almost all redevelopment initiatives, whether from the private sector for clearly speculative reasons, from the public sector with genuine concern for the housing and living conditions of the poor, or even when tenants have organized themselves to promote redevelopment (UNCHS 1984b).

An example of a more successful approach is Bangkok's land-sharing scheme, an interesting, realistic approach in a country where an open market system predominates (UNCHS 1984b). Land sharing works by partitioning plots occupied by slum dwellers into two parts—one for rehousing the dwellers and one for the landowner to develop. Land sharing is thus a compromise solution, a negotiated agreement (UNCHS 1986). But only five among Bangkok's many low-income communities that had been served with eviction orders won their right to stay (UNCHS 1986). Further, land sharing, with its implicit potential in conflict resolution, has limited application to most inner cities. Many are already densely populated, making further increases in density impossible without building costly high-rise apartments.

Sites-and-services and area-upgrading programs have been an unquestionable improvement over the public shelter programs that preceded them. They usually changed the outlook of policymakers and public agencies, convincing them that heavily subsidized, high-standard, high-cost units could only be good-looking social failures (Renaud 1987). The new types of projects played a key role in exposing such important constraints on housing market performance as prohibitively costly building regulations and overlapping layers of traditional and modern property rights, further complicated by imported land use laws or other regulatory constraints.

In evaluating these projects, it is important to remember their funding is typically from bilateral or international agencies; access to foreign capital facilitated the emphasis on the affordability, design, and construction of low-income housing units. But the total numbers reached worldwide are small compared to demand. Indeed, total capital funds available from all international lenders would not be enough to finance urban investment needs in a single large country such as Brazil or India (Renaud 1987). The relative importance of these types of projects has declined from about 42 percent of the total volume of operations in the early 1970s to less than 8 percent in the late 1980s (World Bank 1989a). And there are major regional variations in the relevance of sites-and-services projects. In Africa, they have accounted for over 20 percent of total shelter costs; in Latin America, less than 7 percent.

Enabling Strategies

A new strategy—commonly referred to as an *enabling strategy*—is centered around the idea that governments should serve as enablers in the housing sector, withdrawing from their role as housing providers and playing a more forceful role in facilitating new private-sector construction by creating an appropriate regulatory environment and ensuring availability of finance (World Bank 1988a). It is generally accepted that governments do not usually respond to demand faster or more efficiently than private markets (Mayo, Malpezzi, and Gross 1986).

Bangkok provides an example of a strong formal, corporate private sector that has responded rapidly to changes in the housing market. Bangkok's private developers did not enter the market until the late 1960s, but they made rapid inroads. In the early 1980s, when demand for middle-income housing began to weaken, developers began building condominiums, shop-housing, and terraced housing (townhouses). As the market for these units became saturated, developers then turned to inner-city redevelopment projects. Bangkok's problem is that developers have no incentive to provide low-income housing.

In addition to providing an appropriate environment to induce corporate developers to build low-income housing, governments need to aid the efforts of the informal and individual private sectors. Incrementally built housing is a more effective means to increase the volume and improve the quality of shelter than public housing. Such investments in self-help housing make particular sense in Third World cities because they demand less of such scarce resources as high-skilled labor, capital, and foreign exchange, using instead relatively abundant production factors (Doebele 1987). Given most developing countries' acute resource constraints, it is now being accepted that incremental building is the functional substitute for the incremental paying that takes place in countries with mortgage systems that reach the working class. To facilitate incremental building by the poor, local governments must alter building codes and regulations and stimulate appropriate technologies by providing incentives to small-scale producers of local construction materials.

Housing Finance

An important area in which governments need to become more active is housing finance. As Renaud (1987) noted, few aspects of economic development remain as unexplored and poorly analyzed as housing's potential to induce financial development and ways to improve it. Housing occupies a very low place in growth strategies. Yet potential efficiency gains to be derived from well-run and decentralized housing finance systems appear to be large, because housing investment represents 12-25 percent of total annual investment in most countries, although mortgage financing covers less than 10 percent of annual transactions in most developing countries (Renaud 1987).

In a majority of cities, low-income households rely primarily on informal financing methods (e.g., borrowing from friends or family members). In Cairo, much of the financing for housing comes from earnings of workers abroad, sales of inherited land or jewelry, and savings in informal credit associations. For the majority, formal housing finance is irrelevant. A prospective homeowner must have legal title to the land, obtain a building permit, and use formal construction methods to be eligible for credit (UNDIESA 1990a). Low-income households find the transition to home ownership in Cairo increasingly difficult because costs are rising more rapidly than incomes. Households without access to repatriations find it even more difficult to compete.

Few developing countries have successful housing finance systems. During the 1970s, when inflation was rapid in most of these countries, many housing finance institutions lent at negative real rates of interest that often led to considerable decapitalization by the early 1980s. The certain consequence of keeping mortgages below market rates is that they are rationed, usually to benefit those perceived to have the lowest risk of default—wealthier people or those government policy favors, such as civil servants (Mayo, Malpezzi, and Gross 1986).

Conclusion

Most Third World cities will experience dramatic population growth over the next several decades, resulting in tremendous pressures on land and housing markets. The combined forces of population growth and suburbanization have used much of the accessible land, and the period of easy access to urban land is now over. Access to land has been hindered by the physical growth of cities and inadequate development of transport services. In most cities, peripheral land is now far from the city center. Work in the central city is generally available only to persons prepared to spend several hours each day commuting.

Mechanisms to influence public control of the land market in Third World cities are not difficult to suggest; the major problem is implementation. Some policies may not be appropriate in a particular city because of institutional constraints (e.g., land nationalization in a market-oriented economy). It is not politically easy to introduce effective policies if powerful vested interests are determined to maintain the status quo.

Despite new technology that provides an excellent means of modernizing cadastral and land registration systems, the real reason why so many cities have such antiquated systems and why land and housing values have not been updated is that owners do not wish to pay taxes. Governments are highly sensitive to the complaints of landowners. Clearly, without the political will, cadastral systems will not be improved. Likewise, the whole issue of land speculation comes down to a matter of political will and competent state management. Both are in question in most Third World cities (Gilbert 1990).

Regarding housing, approaches to interventions have evolved considerably since the 1950s from a public housing/redevelopment approach, through a "basic needs" emphasis, to the present "enabling" approach that recognizes the limitations of direct government intervention and emphasizes indirect public-sector support through provision of a policy and regulatory framework. Clearly, we now need more emphasis on land assemblage, infrastructure investments, the creation of specialized national financial systems for shelter, and the encouragement of small builders and suppliers of indigenous building materials.

Many developing countries are ill prepared for the coming dramatic expansion of their cities. It is not certain that even the limited progress of the past will be sustained in the future. The problems of land and housing are so severe that it is futile to pretend they can be solved. They can only be reduced.

5

Job Creation Needs in Third World Cities

DENNIS A. RONDINELLI
JOHN D. KASARDA

New economic challenges are arising in developing countries from rapid labor force growth, redistribution of people from rural to urban areas, and the shift of labor from agricultural to nonagricultural occupations. The International Labour Office (ILO 1986) estimates that over the next 35 years the economically active population in developing countries will increase by about 1.2 billion. Most of this increase will be in urban areas. As a result, shortly after the turn of the century, for the first time in history, more than half of the population in developing countries will live in urban areas, and over half of the economically active population will be nonagricultural (UNDIESA 1988b).

Generating sufficient numbers of jobs to absorb the rapidly growing labor force in developing countries will require substantial investments. In the Caribbean Basin alone it is estimated that between 1980 and 2025 the labor force will triple from 53 to 150 million (Espenshade 1988). Nearly 1 million new jobs will have to be created annually in Mexico alone, at a cost of about $102 billion (in 1982 dollars), to absorb increases until the year 2010 in its economically active population. Central America and the Caribbean will need to generate about more than 5 million new jobs just to keep unemployment rates at 1990 levels, requiring investments of $146 billion.

If the public and private sectors in developing countries cannot expand investment, production, and employment opportunities to keep pace with the growth of the economically active population—especially in urban areas, where the labor force will be growing fastest—these demographic trends will have serious adverse effects on developing countries' ability over the next two decades to achieve better standards of living.

Although much research has been done on population growth, migration, and urbanization in developing countries, little of it has examined the changes in the size, characteristics, and location of the economically active population. Moreover, little applied research has been done on the implications of these demographic changes for job creation needs in urban areas. Whereas governments of developing countries have been experimenting with policies to redistribute population and expand employment opportunities, there have been few systematic evaluations of the results (UNDIESA 1989c).

In this chapter we examine four components of the job creation issue. We describe a conceptual framework for assessing the scope and dimensions of demographic and employment problems in developing countries. We examine the trends in urbanization and urban labor force growth in Africa, Asia, and Latin America and identify the principal policies and programs that governments in developing countries have used to expand employment opportunities in urban areas. Finally, we explore the research implications of these trends.

A Conceptual Framework for Analysis

Policies for dealing with employment needs in urban areas of developing countries must be based on assessments of the relationships among overall demographic shifts; changes in the size, characteristics, and location of the economically active population; urban job creation needs; and public- and private-sector policies for job creation.

Our conceptual framework (Figure 5.1) postulates that demographic changes, spatial redistribution of population, and changes in the occupational and industrial characteristics of the labor force in developing countries will result in dramatic changes in the size, characteristics, and location of the economically active population. Anticipated sluggish growth in agriculture and manufacturing in most developing countries, and in formal-sector employment opportunities generally, will leave these countries with huge employment deficits in urban areas over the next 35 years. The ability of developing countries to absorb their growing urban labor forces will depend on the successful implementation of public policies

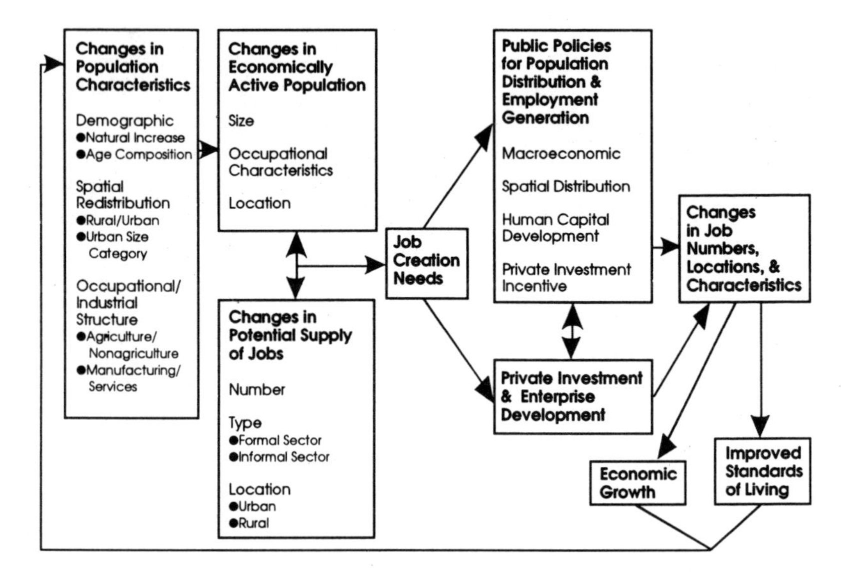

Figure 5.1. Conceptual Framework for Assessing Job Creation Needs and Policy in Developing Countries.

that stimulate national economic growth and private-sector investment in activities that generate new employment opportunities.

The success of economic growth policies and the expansion of private enterprise will determine the number, types, and location of jobs in developing countries over the next three decades. The ability of government and private enterprise to expand employment opportunities in urban areas will determine the capacity of developing countries to achieve sustainable economic growth and to improve the standard of living of their populations. Here we focus on each of these components of the conceptual framework to explore problems of urbanization and employment in developing countries.

Labor Force Growth in Developing Countries

Three types of demographic changes have an impact on the characteristics and size of the urban economically active population: overall population growth rates, spatial redistribution of the population, and changes in occupational and industrial structure of the labor force.

Population Growth Rates and Characteristics

The rapid growth of the urban labor force in developing countries results from high rates of overall population growth and the large proportion of the population in young age groups. Although progress has been made over the past two decades in lowering population growth rates in many developing countries, population will continue to grow on average at more than 2 percent a year over the next decade (UNDIESA 1989c). Thus, the total number of Third World people will increase by 3 billion over the next 35 years. The population of developing countries rose from 2.6 to 3.6 billion between 1970 and 1985 and is projected to grow from about 4 billion to more than 7 billion between 1990 and 2025 (UNDIESA 1989c). In 1990, about 77 percent of the world's population was living in developing countries, but the proportion will climb to 84 percent by 2025.

Another demographic factor affecting labor force characteristics is age structure. Because many developing countries have high rates of population growth, large proportions of their populations are in the dependent age groups. In 1980, over a third of the labor force in developing countries was under age 24 and nearly half was under 30 (ILO 1986). The ILO estimates that in 1990 about 46 percent of the people in developing countries were under 20 years old, and the median age was 22. With such a large portion of the population in young age groups, growth of the labor force will continue well into the next century.

Developing countries also have relatively high labor participation rates. About 94 percent of the men aged 20-59 are economically active in developing countries, as are 53 percent of women aged 20-49. Young people in developing countries enter the labor force much earlier in life than do young people in more developed countries. About 60 percent of males and 41 percent of females aged 15-19 are economically active in developing countries (UNDIESA 1988b). About 15 percent of children between ages 10 and 14 are officially recorded as members of the labor force, but in poor countries and rural areas far more children make an economic contribution to household income.

Spatial Redistribution of Population

The need for jobs in urban areas will increase not only as a result of rapid population growth, but also from a massive spatial redistribution of population (UNDIESA 1989c). For nearly 40 years, urban population in developing countries has grown by more than 3 percent a year, a rate much higher than total population growth.

By early in the next century, population living in urban places in developing countries will exceed that of rural places. More than half of East Asians will be urban dwellers. About 65 percent of the Caribbean population will be urban, as will nearly 71 percent of Central American and almost 81 percent of South American populations (UNDIESA 1989c). Of the 3 billion increase in population expected in developing countries between 1990 and 2025, 87 percent will live in urban places.

Because these high rates of urban population growth will continue during the 1990s, the number of people living in cities and towns by the turn of the century will reach nearly 2 billion, double the 1980 number. In the following 25 years, the number of urban dwellers will double again from 2 to 4 billion.

Urban population is growing in all city-size categories in most regions of the developing world, and even the largest cities will continue to see substantial population growth. In 1950, 31 cities had more than 1 million residents. By 2000, such cities will number about 300. By the end of the 1990s, 20 of the world's 25 largest cities will be in developing countries (UNCHS 1987).

Growth of the Economically Active Population

The factor most directly affecting job creation needs in most developing countries is the rapid growth of the economically active population. Unfortunately, understanding the scope and magnitude of labor force growth is constrained by weak data bases. Often, labor surveys in developing countries are poorly designed, cover only formal-sector wage labor, or are not comparable from one survey period to another. Different countries use different definitions of occupational categories and of employment and unemployment.

Despite these limitations, the data available provide information on "order of magnitude" changes in labor force characteristics in developing countries. Clearly, between 1950 and 1990 the world's labor force increased by a little more than 1.2 billion people. Between 1990 and 2025, it is projected to rise from 2.4 to about 3.6 billion (Table 5.1). Nearly 96 percent of the increase in the world's labor force will take place in developing countries.

In Africa, where the economically active population more than doubled between 1950 and 1990, the labor force will grow from about 243

Table 5.1 World Labor Force Growth

Location	Economically Active Population (millions)								
	1950	*1960*	*1970*	*1980*	*1990*	*2000*	*2010*	*2020*	*2025*
World	1189.9	1347.4	1596.7	1957.3	2363.5	2752.5	3146.7	3498.3	3648.9
More Developed Regions	386.7	431.2	476.9	541.8	585.5	614.8	635.9	635.5	636.3
Less Developed Regions	802.2	916.2	1119.8	1415.5	1778.0	2137.7	2510.8	2862.8	3012.6
Major Geographical Areas									
North America	70.6	79.9	95.9	121.8	135.4	146.6	155.7	157.8	158.2
Western Europe	135.2	142.3	149.7	161.5	172.4	175.3	174.0	168.2	163.6
Eastern Europe	46.1	49.2	54.2	56.4	59.3	62.9	65.1	65.6	66.1
ex-USSR	93.9	110.0	117.3	136.9	146.6	154.9	167.2	170.9	175.3
Africa	99.2	119.4	147.6	189.2	242.8	318.3	425.4	565.9	649.6
Latin America and Caribbean	128.5	151.2	186.8	245.3	293.7	346.5	401.9	446.7	465.8
Asia	643.8	724.2	883.6	1100.7	1374.3	1616.2	1835.3	2003.4	2050.4

SOURCE: UNDIESA 1988b.

million in 1990 to nearly 650 million by 2025. In Latin America and the Caribbean it will increase by nearly 60 percent. The largest increase in the economically active population will be in Asia, where the labor force is expected to grow by more than 670 million between 1990 and 2025, to more than 2 billion people.

Changes in Labor Force Occupational and Industrial Structure

Of particular significance for developing countries is the rapid growth of the economically active population and the dramatic changes expected in the structure of the labor force (Table 5.2).

About 80 percent of the labor force in developing countries was engaged in agriculture during the 1950s. Growth rates of agricultural labor began to decline in the late 1970s, however, even as the number of people in manufacturing and services grew on average by more than 4 percent a year. The United Nations (UNDIESA 1988b) projects continuing growth of nonagricultural labor in developing countries at more than 3 percent a year over the next 35 years. As a result, in 1990 only 59 percent of the labor force in developing countries was in agriculture. Shortly after the turn of the century, more than half will be found in manufacturing and service sectors (Table 5.3).

In many developing countries, these changes in the occupational structure of the labor force will have pervasive economic implications. For example, Indonesia's labor force in manufacturing and service occupations grew by about 4.2 percent a year during the late 1980s and is expected to continue growing at more than 3.5 percent a year during the 1990s. The number of Indonesian nonagricultural laborers more than doubled from 15 to 36 million between 1970 and 1990. It is projected to more than double again over the next 35 years, reaching 94 million. Thus nearly 60 million new nonagricultural jobs must be created during this period if unemployment and underemployment rates are not to rise. Most of the jobs needed in nonagricultural activities will have to be in urban areas; Indonesia will have to create more *new* off-farm jobs over the next three and a half decades than the *total* number of jobs that existed in the country through the mid-1980s.

By 2025, more than two-thirds of the labor force in developing countries will have to find jobs in manufacturing and services. In Latin America and the Caribbean more than 90 percent of the labor force will be in those sectors, as will be 62 percent in Asia and 56 percent in Africa. Not all nonagricultural workers will live in urban areas, but in

Table 5.2 Growth Rate of Economically Active Population in Agricultural and Nonagricultural Occupations

	Average Annual Growth Rate (percentages)							
Location	*1950–1955*	*1960–1965*	*1970–1975*	*1980–1985*	*1990–1995*	*2000–2005*	*2010–2015*	*2020–2025*
More Developed Countries								
Total Labor Force	1.22	1.08	1.28	0.94	0.49	0.44	0.03	0.02
Agricultural	–1.36	–2.91	–1.97	–3.22	–3.74	–3.75	–4.12	–4.04
Nonagricultural	2.67	2.49	1.95	1.49	0.84	0.66	0.17	0.11
Less Developed Countries								
Total Labor Force	1.19	1.89	2.99	2.43	1.96	1.76	1.69	1.45
Agricultural	0.53	1.21	1.51	1.47	0.79	0.52	0.38	0.06
Nonagricultural	3.76	3.88	4.12	4.14	3.54	3.25	3.08	2.57

SOURCE: UNDIESA 1988b.

Table 5.3 Size and Percentage of Labor Force in Nonagricultural Occupations in Less Developed Countries

Location	*1950*	*1960*	*1970*	*1980*	*1990*	*2000*	*2010*	*2020*	*2025*
Number (millions)									
Less Developed Countries	154.4	223.1	327.8	490.7	725.4	1013.2	1354.0	1715.1	1887.4
Africa	18.0	25.7	37.8	59.2	89.4	135.3	204.6	300.9	360.7
Latin America & Caribbean	88.5	115.5	145.5	201.6	249.6	301.9	358.4	406.0	427.0
Asia (excluding Japan)	109.2	159.8	235.5	346.5	518.0	718.5	942.5	1161.5	1253.1
Percentages									
Less Developed Countries	19.3	24.4	29.3	34.7	40.8	47.4	53.9	59.9	62.6
Africa	18.2	21.6	25.6	31.3	36.8	42.5	48.1	53.2	55.5
Latin America & Caribbean	68.9	73.8	77.9	82.2	84.8	87.1	89.2	90.0	91.7
Asia (excluding Japan)	18.8	24.7	29.8	34.3	40.1	46.4	52.9	59.2	62.2

SOURCE: UNDIESA 1988b.

many developing countries even rural and periurban residents will have to find full- or part-time off-farm work in small towns and secondary cities (Rondinelli 1983).

Rapid Growth of the Urban Labor Force

Few reliable projections have been made of the urban-rural distribution of the economically active population. If we assume that the labor force will be distributed between rural and urban areas in about the same proportion as the general population, however, we can conservatively estimate that the urban labor force will double from about 409 million in 1980 to more than 839 million in 2000 (UNDIESA 1988b). It will double again to 1.7 billion by 2025 (Table 5.4).

This assumption may underestimate the real size of the urban labor force because it does not account for large numbers of rural people who either commute to urban jobs or spend agricultural slack seasons working in towns and cities. Many rural households have one or two members who work part of the year in urban areas and part of the year on the farm. If the definition of the urban labor force is expanded to include temporary urban workers and commuters who live in rural areas, its growth will be higher than estimates based on residential location of the population.

This rationale implies that more than 228 million new jobs must be created in urban areas in developing countries during the 1990s alone. An additional 875 million urban jobs will be needed in the following 25 years. To absorb the enormous growth of the urban labor force, at least 280 million new urban jobs will be needed in Africa, 180 million in Latin America and the Caribbean, and almost 672 million in Asia.

The rapid growth of the urban labor force over the past 20 years in many developing countries can be largely attributed to migration. In Asia, for example, studies by the United Nations Economic and Social Commission on Asia and the Pacific studies (UNESCAP 1988) attribute about 60 percent of the 1975-1980 increase in Bangkok's labor force to migration. About 62 percent of the rise in South Korea's urban labor force then was attributed to net in-migration. In the future, the large populations in many Third World metropolises and trends in permanent and temporary migration and natural increase are all likely to be significant in urban labor force growth.

Table 5.4 Estimated Urban Labor Force Growth in Less Developed Countries, 1990-2025

	1990	*2000*	*2010*	*2020*	*2025*
Less Developed Countries					
Economically Active Population (millions)	1778.0	2137.7	2510.8	2862.8	3012.6
Population in Urban Areas (%)	33.6	39.3	46.2	53.1	56.5
Urban Labor Force (millions)	597.6	839.7	1158.7	1521.0	1701.2
Africa					
Economically Active Population (millions)	242.8	318.3	425.4	565.9	649.6
Population in Urban Areas (%)	32.7	39.1	45.7	52.3	55.4
Urban Labor Force (millions)	79.4	124.5	194.4	295.9	359.9
Latin America & Caribbean					
Economically Active Population (millions)	293.7	346.5	401.9	446.7	465.8
Population in Urban Areas (%)	72.0	76.8	80.1	82.9	84.1
Urban Labor Force (millions)	211.5	266.1	321.9	370.3	391.7
Asia					
Economically Active Population (millions)	1374.3	1616.2	1835.3	2003.4	2050.4
Population in Urban Areas (%)	29.9	34.9	41.9	49.3	52.8
Urban Labor Force (millions)	410.9	564.1	768.9	987.7	1082.6

SOURCE: UNDIESA 1988b.
NOTE: Labor force assumed to be distributed in proportion to urban-rural population distribution.

The Supply of Jobs in Urban Areas

The enormous number of people being added to the urban labor force in developing countries is placing severe strains on the ability of private and public organizations to increase production and expand employment opportunities sufficiently to absorb the growing numbers of people looking for work. Unless productive capacity can be expanded rapidly enough to create about 1 billion new jobs in urban areas over the next 35 years, widespread urban unemployment and underemployment will undermine progress toward economic growth and improved living conditions.

Few developing countries have been able to increase production at a rate equal to or higher than urban labor force growth. Even developing countries that have experienced relatively strong economic growth faced serious problems in providing enough jobs. Surveys of the seven largest urban areas in Colombia, for example, showed an increase in labor force size from 1.5 million people in 1951 to 7.3 million in 1984. Urban labor force growth increased from about 4.6 percent annually during the 1950s to 5.2 percent a year in the 1970s and 1980s. Even with employment growth averaging 4.4 percent a year, the unemployed population grew from about 312,000 in 1976 to 600,000 in 1985. Unemployment in these Colombian cities averaged 8.7 percent from 1976 to 1982 and then soared to 14 percent in 1985 (deGomez, Ramirez, and Reyes 1988).

Although several Asian countries increased per capita real gross national product (GNP) by more than 3 percent a year between 1965 and 1985, per capita GNP increased by only 2.5 percent a year in Latin America and by 1.1 percent a year in Africa. Many low-income African countries had a negative growth rate (Tucker and Treado 1988, Table D-1).

Sluggish growth rates combined with rapid expansion of the labor force have led to employment problems in urban areas in many developing countries. Not surprisingly, the most severe urban job creation problems are in the poorest developing countries. In urban areas in sub-Saharan Africa, regular wage employment typically provides less than 10 percent of total employment. Opportunities for regular employment in urban areas have fallen over the past decade in many African countries, and in others they have lagged behind the pace of growth of the nonagricultural labor force. In Kenya, Malawi, Mauritius, Zambia, and Zimbabwe, the nonagricultural labor force grew much faster than did urban wage employment during the 1980s (van Ginneken 1988).

Unemployment and Underemployment

Little accurate information is available on open unemployment in urban areas. As one labor expert pointed out, "The unstructured nature of labor markets in low-income developing countries causes unemployment figures to greatly understate their joblessness problems" (Marshall, 1988, p. 169). But recent studies indicate that unemployment in most Latin American cities increased during the 1980s. Unemployment rose by more than 50 percent in Bolivia, Colombia, and Peru. In 1984, open unemployment in urban areas exceeded 13 percent in Bolivia, Chile, Colombia, Uruguay, and Venezuela; it was greater than 7 percent in Brazil and Costa Rica (International Labour Organisation 1986a). Unemployment also increased substantially in many Caribbean countries during the 1960s and 1970s—for example, from about 8 to nearly 14 percent in Barbados, 8-20 percent in Guyana, and 13-27 percent in Jamaica—and remained relatively high during the 1980s (Hope 1984). In many African countries, the open unemployment rates are higher than in Latin America and the Caribbean.

Underemployment is an even more serious problem. As much as 40 percent of the urban work force in many Third World cities is underemployed (ILO 1987). At the same time, expansion of urban employment is highly variable. Jobs have not been created fast enough in the urban areas of most developing countries to absorb the growing working-age population. For example, in Madras, India, employment grew during the 1970s by only about 1.7 percent a year even as the population was growing by a 3 percent rate (UNDIESA 1987c). In Bangkok, the 1980 census showed open unemployment rates of about 7.5 percent, but service, sales, and clerical workers experienced average unemployment rates of nearly 15 percent, and about 30 percent of the city's craftsmen, production workers, and laborers were unemployed (UNDIESA 1987a).

Problems also arise from the spatial distribution of jobs. In many of the poorest countries, jobs in manufacturing and some commercial services are highly concentrated in one, or a few, major cities. Nearly half of Bangladesh's total formal-sector manufacturing employment is found in Dhaka. Dhaka accounts for nearly all of the country's employment in rubber products, 97 percent of jobs in furniture making, 84 percent of footwear production, 82 percent in leather goods, and more than half in machinery production and textiles (UNDIESA 1987b). Cairo has over half of Egypt's skilled tradesmen and workers in transformative industries, and 36 percent of its production workers (Khalefa and Mohieddin 1988).

Job Creation in Urban Areas

Poor countries with rapidly growing urban populations face the most severe problems in expanding production and employment. But even countries where population growth is declining will urgently need to create new jobs during the 1990s. In Indonesia, for example, a successful family planning program has reduced the overall population growth rate to about 2 percent. Nevertheless, Indonesia's labor force is growing at about 2.6 percent a year as those born in the 1970s enter the labor force. About 1.8 million new jobs will be needed each year through the end of the century. In the 1990s, 30 percent more jobs will have to be created every year than were created each year during the 1970s when Indonesia's economy expanded most rapidly (Kingsley 1988). Little accurate information on job creation exists for most African countries, but studies of Lagos, Nigeria, show the labor force grew from about 1.6 million people in 1978 to about 2.8 million in 1985, about twice as fast as the rate of job expansion (McNulty and Adalemo 1988).

Job Creation in Manufacturing. Although large-scale manufacturing firms have created an impressive number of jobs in the newly industrializing countries of Asia and Latin America, they have generated only a relatively small number of jobs in most other developing countries. Larger manufacturing establishments are unlikely to expand fast enough to absorb more than a small portion of the rapidly growing urban labor force in the years ahead. For example, in Asia the share of the labor force employed in manufacturing increased by only 5 percent in South Korea and Malaysia (from 19 to 24 percent and 12 to 17 percent, respectively) between 1975 and 1985, but only by about 1 percent (from about 8 to 9 percent) in Indonesia and Burma (Edgren and Muqtada 1989). The share of the labor force in manufacturing declined in Hong Kong, Singapore, the Philippines, Thailand, and Pakistan.

This is not to say that large-scale manufacturing has no role to play in urban job creation. In Latin American cities like Guadalajara, Mexico, large industries have been important sources of direct and indirect employment. In addition to providing jobs in factories, large industries subcontracted to small-scale and informal-sector enterprises that employed owners, their families, and wage laborers (Roberts 1989). Nonetheless, most of the jobs being created in urban areas of developing countries are in commerce and services rather than in manufacturing and industry. Even in the newly industrializing Asian countries, the service sector absorbed the largest portion of the employed labor force—

more than 40 percent in Malaysia, about 50 percent in South Korea, 62 percent in Hong Kong, and 73 percent in Singapore—from the mid-1970s to the mid-1980s. In the poorest Asian countries, a large majority of nonagricultural laborers work in services—76 percent in Bangladesh, 71 percent in Pakistan, 78 percent in Indonesia, and 79 percent in the Philippines (Edgren and Muqtada 1989).

Job Creation in Informal-Sector Commerce and Services. Except for some of Asia's newly industrializing countries, in almost all regions of the developing world the majority of the urban labor force works in such informal jobs as hawking and vending, cottage industry, and small or microsized enterprises or by providing low-cost services. Informal-sector enterprises—small-scale, mostly family-operated or individual activities that are not legally registered and usually do not provide employees with social security or legal protection—absorb more than half of the urban labor force in many large cities and 70-90 percent in secondary cities and small towns (Rondinelli 1983).

The informal sector is a major source of jobs for the urban population even in countries where manufacturing has expanded over the past half century. Despite rapid industrialization in many Latin American countries from the 1950s to 1980, the portion of the labor force employed in informal-sector activities remained almost the same. It declined less than 0.05 percent and still constitutes about 30 percent of the economically active population in Latin America. If the definition of informal workers includes wage laborers in small-scale enterprises that offer no social security or other form of labor protection, the portion of the labor force in the informal sector increases to about 50 percent (Castells and Portes 1989).

Studies estimate that about 46 percent of the Bogotá, Colombia, labor force is in unprotected jobs. Those employed in the informal sector tend to be people in dependent age groups, the poorly educated, and recent migrants. Labor force participants under age 19 and over 55 are heavily represented in Bogotá's informal sector. Half of the informally employed workers have completed only elementary school. Recent migrants to Bogotá are more heavily represented than long-term residents. Nearly all of the informally employed earned less than the minimum wage (dePardo and Castano 1989).

Even in Asia, where growth in manufacturing and service jobs has been relatively high, the "organized" formal sector has absorbed only a small percentage of the annual labor force growth: About 5 percent in Bangladesh and 10 percent in India. In South Asian countries—India,

Bangladesh, and Pakistan—at least 80 percent of all manufacturing employment is accounted for by small-scale and cottage industries. In the poorest countries, like Nepal, they account for as much as 96 percent (International Labour Organisation 1987).

In urban Indonesia, the informal sector accounts for more than half of those employed in nonagricultural jobs. Over 3.5 million urban Indonesians worked in the informal sector in 1980. Nearly 55 percent of them were employed in trade and services and only 8 percent in manufacturing (Sethuraman 1985). Similarly, in Dhaka, Bangladesh, about 65 percent of the city's employment is in the informal sector. Over half are self-employed, about 25 percent are hired labor, and about 19 percent are family labor (Amin 1987).

In secondary cities and towns, small tertiary-sector enterprises almost wholly dominate the economic structure. Most secondary cities, except in a few newly industrializing countries, have a fairly small share of the country's manufacturing activities. In Peru's intermediate city of Huancayo, Long and Roberts (1984) found about half the male workers were self-employed or family workers and 76 percent worked in small-scale and informal-sector enterprises. The reasons are not difficult to discern. Roberts (1989, 41) points out:

> The advantages of informality for generating employment were obvious. Overhead costs were low when compared with the tax and social security obligations of formally registered firms; it was easy to become a small scale entrepreneur, requiring no more than space in a house and a small amount of capital in tools, materials or articles for sale; family relationships provided flexible access to labor, to credit and to economic information.

Surveys of four secondary cities and towns in India—Wardha, Ghaziabad, Allahabad, and Jaipur—ranging from a little more than 80,000 to 1 million in population showed surprisingly similar economic characteristics (National Institute of Urban Affairs 1987). Over 95 percent of all enterprises in these cities were informal-sector ones. In Allahabad 53 percent of the labor force and in Wardha 63 percent were employed in the informal sector. Over half of these enterprises involved retail trade or commercial activities, mostly employing family members and occasionally one or two hired workers. Manufacturing, services, and repair enterprises employed on average two to five employees and construction enterprises from six to nine paid workers. On average, the employment-generating capacity of informal-sector enterprises was low. The

2,000 enterprises surveyed in the four cities had only 3,141 workers—self-employed, casual, contract, apprentice, and unpaid family workers.

Thus, in large cities and small towns, the informal sector remains the dominant employer. Studies of informal-sector enterprises in Jamaica indicate their remarkable ability to survive precisely because of their small size and flexibility. In Jamaica "small businesses are often available to customers at all times, and are willing to produce upon demand. They carry little in the way of inventory and much work is custom or 'job' in nature" (Doeringer 1988, 467). Flexible labor practices are an advantage because owners' and family members' labor can be tapped quickly; they are willing to work long hours and carry few additional costs when they are not working.

Not much attention has been paid to characteristics of informal-sector enterprises in cities of different sizes, but some evidence from Kenya suggests that those in large metropolitan areas may differ from those in smaller cities and towns. Hosier's studies in Nairobi (1987), with a population of more than 1 million, and in Meru, with less than 100,000, indicated that Nairobi's informal enterprises made greater use of nonbinding employment commitments by using a larger portion of casual and family labor than did enterprises in Meru. Small-scale enterprises in Meru did more custom work, had stronger clientele relationships, and tended to earn more income per enterprise than did those in Nairobi. In smaller towns, enterprises could establish closer and more stable clientele networks.

In sum, developing countries everywhere face serious challenges of expanding urban employment opportunities and of absorbing the large numbers of people entering the labor force over the next 35 years. In most countries, large-scale formal-sector manufacturing firms can provide jobs for only a small portion of people entering the urban labor force. Small- and medium-scale enterprises in manufacturing, services and trade, and microenterprises in the informal sector will continue to be the major sources of jobs.

Policies for Population Distribution and Job Creation

Governments in developing regions have been experimenting with policies to influence the distribution of population and the expansion of employment opportunities for nearly 40 years. Four types of policies have been adopted in most developing countries: (a) macroeconomic adjustment policies aimed at creating a conducive environment for employment growth, (b) spatial redistribution policies to prevent overconcentration of the economically active population in the largest cities, (c) human resource development policies to prepare the labor force better for modern sector employment, and (d) incentive programs to stimulate private investment and develop private enterprise in high-priority locations.

Macroeconomic Policies Affecting Employment

Most developing countries have attempted to stimulate job creation through macroeconomic adjustment policies. During the 1950s and 1960s, governments in most developing countries adopted import substitution strategies for accelerating industrialization and employment growth.

For example, all of East and Southeast Asia, except Singapore, sought to industrialize by promoting local production of imported capital and consumption goods and protecting local industries from international competition. The strategy initially spurred economic growth and employment but quickly lost momentum as a result of two crucial bottlenecks—limited domestic markets for industrial goods and difficulties of expanding into intermediate goods industries. Reliance on protectionist policies reduced international competitiveness. In the early 1970s, South Korea, Hong Kong, and Taiwan switched to a Japanese-style export industries development strategy while maintaining labor-intensive production and protecting domestic industries in which they had competitive advantages. During the late 1970s and early 1980s, Malaysia, the Philippines, Thailand, and Indonesia also switched to export-promotion strategies, and they, too, continued to protect industries producing goods aimed at domestic markets (Chowdhury, Kirkpatrick, and Islam 1988).

Countries like Japan, South Korea, and Taiwan that "have focused strongly on labor intensive measures—whether in agriculture, import substitution industries or the export led phase of their growth—have attained near full employment without sacrificing overall growth" (Edgren and Muqtada 1989, 64).

East Asian countries were successful in expanding employment opportunities through an outward-looking industrial development strategy, but their success depended also on policies promoting an equitable income distribution. They redistributed productive assets through land reform and agricultural development policies before diversifying into labor-intensive manufacturing. They took deliberate steps to broaden domestic markets and effective demand for manufactured goods at home as they developed export markets.

Many African and Latin American countries have been less successful at generating jobs through export industrialization. They tend to focus on large-scale capital-intensive manufacturing, natural resource exploitation, or estate-based agricultural exports without building an equitable foundation for widespread employment and income distribution and to specialize in one or two products subject to unstable fluctuations in international demand.

The policies pursued by many African and Latin American governments to accelerate industrialization, including trade regulation, creation of investment incentives, and direct public investment in manufacturing, favor large firms and discriminate against small- and medium-scale enterprises that generate the most urban employment opportunities (Little 1987).

In the wake of the international "oil shock," serious economic recessions during the late 1970s, and the debt crises of the early 1980s, the International Monetary Fund pressured many governments to restructure their economies. These structural adjustment policies required governments in developing countries to adopt economic stabilization measures aimed at controlling inflation, loss of foreign exchange reserves, capital flight, and public-sector budget deficits. Structural adjustment policies sought to remove obstacles to long-term economic growth by liberalizing market restrictions, eliminating excessive taxes and subsidies, controlling prices and interest rates, reducing high tariffs and import restrictions, and modifying or eliminating distortions in incentives for private-sector investment. The World Bank (1983a) provided structural adjustment loans aimed at reducing government employment and public expenditures, privatizing public services, and reducing government interference in market activities (Khan 1987; Lal 1987).

However badly needed, the impacts of macroeconomic adjustment policies on job creation have been mixed. In many developing countries, the policies have resulted in higher urban unemployment and lower incomes for poorer segments of the urban labor force. In countries like Costa Rica, where structural adjustment policies were quite successful in restoring an economic environment conducive to export promotion and private-sector investment, unemployment eventually dropped, incomes rose, and poverty decreased (Fields 1988). But in other Latin American countries, inappropriate or ineffectively implemented policies adversely affected national labor markets (Corbo and de Melo 1987).

Adjustment policies adopted in most Latin American countries were biased against lower-income urban workers, who could least afford the cost of adjustment. According to Tokman (1988, 119):

> This recessive policy package generated a contraction in the demand for labor by modern firms which, given the rapidly increasing labor supply, produced a significant expansion of open unemployment. Not only were fewer jobs created, but those that were generated were of lower quality. A rapid increase in informal employment, in services compared to manufacturing, and in public employment in relation to the modern private sector diminished average productivity and resulted in a deterioration in income.

Manufacturing employment in Latin America declined on average by 2.2 percent a year between 1980 and 1985 even as wage percentages declined in construction (–3.3), industry (–2.3), and in the public sector (–3.2).

The poorest urban laborers in Asia were adversely affected by the impact of recession and lower rates of economic growth, then by the impacts of structural adjustment policies. The ILO pointed out that structural adjustment policies in many Asian countries left the poor worse off because they reduced incomes, increased unemployment, and raised the prices of food and other essentials: "In the worst affected country, the Philippines, large numbers of people, including modern-sector wage employees have been exposed to loss of real income" (ILO 1987, 69). In most countries, those outside the modern-sector labor market suffered the worst setbacks.

Policies Promoting Spatial Redistribution of Population and Economic Activities

Over the past three decades, governments in most developing countries have attempted to direct migration away from the largest cities to secondary cities and towns and to expand rural and frontier employment opportunities. The strong concentration of urban population and modern productive activities in one or two major metropolitan areas has been a continuing source of dissatisfaction for governments in many of these countries (UNDIESA 1989c).

Governments have tried to moderate the continued concentration of economic activities in metropolitan areas by adopting policies to slow primate city or metropolitan population growth, relocating the national capital, promoting the growth of countermagnet cities, or establishing new towns. Many governments also adopted policies to expand economic activities and employment in small towns and intermediate-sized cities or to promote the expansion of rural "growth centers" (Rondinelli 1988). A large majority of developing countries have experimented with rural development policies, regional development programs for lagging regions and border areas, and colonization schemes aimed at providing job opportunities and slowing migration (UNDIESA 1989c).

In more than 30 countries in Africa, 13 in Asia, and 22 in Latin America and the Caribbean, governments have expressed a desire in their national development plans to limit the growth of their largest metropolitan centers, or at least to slow the pace of expansion (UNDIESA 1989c). Metropolitan growth control policies were never very successful, however, partly because they were never seriously implemented and partly because they operated against fundamental market forces (Dewar, Todes, and Watson 1986; Rondinelli 1990b).

South Korea was one of the few countries that tried seriously to redistribute urban population and economic activities from its primate city to other parts of the country (Rondinelli 1984). During the 1970s, the Korean government sought to slow migration to Seoul and generate employment for people in regions beyond the Seoul metropolitan area. It used a combination of incentives and regulations to control population growth in, and disperse industries from, Seoul. Simultaneously, the government used its investments in overhead capital, social services, physical facilities, and directly productive activities to make secondary cities more attractive for large- and small-scale industries. A complex package of agricultural and rural development policies, price and wage controls, land-use regulations, industrial estate programs, and infrastructure investment and location policies was used to build the capacity of rural towns and intermediate cities to absorb larger numbers of people and support productive activities (Choi 1987; Kim and Donaldson 1979).

The government also tried to create jobs in cities and towns outside the Seoul metropolitan area by making it more difficult for large industries to continue locating there. It enacted stronger zoning regulations, required construction permits for factory building and expansion, and gave financial incentives for industrial relocation. New growth controls included a residence tax on metropolitan citizens, discriminatory tax laws against factories built in the metropolitan area, and discriminatory school fees based on city size (Kim and Donaldson 1979).

The impacts of these policies on slowing Seoul's growth are unclear, however. The land development regulations, combined with strong financial incentives for industrial relocation and substantial government investment in infrastructure and services in secondary and intermediate cities, seem to have been successful in slowing, if not preventing, the continued concentration of people and industries in the national capital and in expanding employment opportunities elsewhere (Rondinelli 1984). But how much of the change in Korea's urban settlement system can be attributed to spatial policies and how much to national economic policies remains unknown.

In other Asian countries, controls on growth of the largest cities were less effective. Experiments with reversing the migration flow and resettling the urban poor in frontier areas or rural colonization projects in Indonesia and Malaysia, restricting migrants' entry to national capitals or limiting their access to urban services in the Philippines and Indonesia, and creating industrial growth poles in rural areas in India and other Asian countries had only marginal effects on slowing growth of the largest metropolises or on generating employment opportunities in other urban places (Simmons 1979).

Governments in developing countries began in the mid-1970s to give greater attention to investing in the services, facilities, and infrastructure needed to make secondary cities more attractive to private-sector investors and rural migrants seeking employment. In Africa, 28 countries claim to have policies for promoting growth of small towns and intermediate cities, 16 say they have growth center strategies, and 7 have created new towns. In Asia, 17 countries claim to have tried small and intermediate city development strategies and 19 have adopted growth center or new towns policies (UNDIESA 1989c).

During the 1980s, nearly all Asian governments sought to improve the capacity of villages and towns to provide basic services and infrastructure in rural areas and to facilitate agricultural production and marketing as a way to slow migration to large cities. Governments in India, Korea, and Malaysia tried to promote towns and small cities as sites for rural industrial and commercial activities that could generate off-farm employment. "New village" and rural colonization projects in frontier areas in Malaysia and Indonesia were used to create settlements for rural migrants (Rondinelli 1991).

Although government policies focused heavily on developing rural towns and small cities, the impact of these policies on redistributing population and generating employment has been variable. Policy implementation has been undermined by lack of political commitment. Malaysia and India, for example, attempted to develop too many small towns and cities at one time, thereby spreading limited financial and managerial resources too thinly (Mahbob 1986). In Pakistan and South Korea, the effects were hindered by lack of cooperation and coordination among national ministries in implementing policies.

Human Resource Development Policies

Most developing countries have adopted policies and programs to enhance human capital as a way to prepare the urban labor force for modern-sector work. Many of these countries have substantial job-training, vocational education, and special education programs, and a few have experimented with relocation incentives and training for segments of the labor force with special skills. Governments in 44 African countries claim to have adopted human resource investment and job-training policies, as have 25 in Latin America and the Caribbean, 24 in Asia, and 10 in the Middle East (UNDIESA 1989c).

The effects of human resource development policies to promote employment have also been variable. One of the keys to the success of newly industrializing Asian countries has been a conscious attempt to develop and use their human capital resources effectively as a part of their economic growth strategies.

An abundant supply of relatively well educated, highly motivated labor was crucial in South Korea's economic growth. South Korea made 6 years of primary education compulsory in 1949 and achieved nearly full primary enrollments by 1960. Government and private firms used various training arrangements to upgrade labor force skills. Vocational high schools were established along with formal and nonformal, institutional, and on-the-job training programs (Golladay and King 1979). The government sought to tailor its human resource development programs to the export-oriented, labor-intensive manufacturing strategy it adopted during the 1970s and 1980s (Park 1988).

Human resource development policies were also crucial to Singapore's successful economic growth and employment expansion strategies. Government expenditures for education grew by 13 percent a year during the 1960s and by more than 30 percent annually in the early 1980s, when Singapore adopted the Basic Education for Skills Training (BEST) program to raise literacy and numeration levels of more than 300,000 workers. Primary and secondary education systems were expanded first; the government then created a system of polytechnic colleges, vocational training institutes, and industrial training centers (Shantakumar 1984). College computer, technology, and science curricula were strengthened, and the National Productivity Board extended programs to improve worker attitudes, productivity, and occupational safety.

Asia's newly industrializing countries were successful in raising the quality of the labor force through human resource development programs at a pace with the expansion of job opportunities and in concert with changes in the industrial structures of their economies. They thereby avoided creating a labor force whose training and skills were inappropriate to economic conditions or generating a large group of educated unemployed—problems that plague many low-income countries in Africa and Latin America.

Incentives to Stimulate Private Investment and Influence the Location of Private Enterprise

Finally, governments in developing countries have provided incentives to stimulate private investment and encourage private enterprise as a way of creating jobs. Most have adopted incentives and controls to influence the location of employment-generating enterprises. Nearly all governments in developing countries provide public infrastructure or subsidize investment in physical facilities to lower production and distribution costs. The United Nations (UNDIESA 1989c) reports that transport rate controls and interregional transportation cost-adjustment policies have been adopted by eight countries in Africa, five in Latin America, and seven in Asia.

About 25 countries in Africa claim to provide grants, loans, and tax incentives to new industries or those relocating to regions promoting job creation. In Latin America and the Caribbean, 18 governments report providing similar incentives, as do 13 in Asia. Further, 9 countries in Africa, 10 in Latin America and the Caribbean, and 17 in Asia and the Middle East report imposing direct restrictions on industrial location, mostly aimed at promoting private investment in rural regions or secondary cities (UNDIESA 1989c).

Socialist and capitalist governments use investment in infrastructure and services to create conditions conducive to employment expansion. From the late 1970s, the government of China, for example, invested in infrastructure, services, and productive activities in three levels of municipalities. Provincial capitals—the secondary cities in the Chinese urban hierarchy—received the bulk of investments for heavy industry, modern infrastructure, heavy utility production facilities, and major highways. Prefectural cities received investments for light industry, agroprocessing, light farm machinery production, and manufacturing plants using intermediate technology and locally available materials. Smaller urban places, the *hsien* cities and rural towns, were seen as centers of rural-urban interaction. They produced small components for manufacturing firms in large cities and farm inputs to rural villages. Investments were made in *hsien* cities for producing energy, cement, fertilizer, and simple farm implements. The government encouraged agricultural equipment repair services and small-scale production of farm inputs. The decisions to make investments in small towns and intermediate-sized cities that were directly related to rural production needs were of critical importance in expanding local employment opportunities (Yu and Gu 1984).

In mixed economies, governments have used subsidies and incentives to influence territorial patterns of private investment and employment expansion, but the impacts have been relatively weak. Government incentives in Indonesia, for example, to entice industries to move out of Jakarta to other cities and to areas outside Java had little influence on industrial location decisions. Firms that moved did so primarily because secondary cities had the services, facilities, and infrastructure they needed to operate profitably (Noer 1985). Incentives for industrial decentralization in Korea simply created a bipolar pattern of urban concentration around Seoul and Pusan, with factories relocating from central Seoul to immediately surrounding districts, redistributing jobs within the Seoul metropolitan area rather than deconcentrating economic activities to other Korean cities (Hahn 1989).

More recently, some Asian countries have taken other approaches to promoting private enterprise development, including privatizing public enterprises and some public services, and encouraging greater private-sector participation in providing urban services and infrastructure. Private-sector provision of shelter, infrastructure, and services is becoming more popular in Asian countries as urbanization creates greater demands on limited government budgets. India's Seventh Plan explicitly recognizes that government alone cannot provide the services and infrastructure needed in urban areas in the future. It calls for private firms and cooperatives to take greater responsibility for investing in housing and other urban services (Sukthankar and Sundaram 1987). Similarly, Thailand's Sixth Plan calls for privatizing many public utilities and urban services (Pakkasem 1987). Malaysia's government is encouraging private-sector participation in land development, transportation, housing, education, health, industrial development, and the financing of urban development projects (Maaruf 1987).

Policy Implications

Because job creation policies to date have had mixed results, governments in developing countries and international assistance organizations must reassess and revise programs. Fields's (1989) review of employment policies in developing countries led him to conclude that in the urban modern sector, employment of unskilled workers has been constrained by labor demand. This implies that governments in developing countries should focus on economic growth policies, especially those that promote exports in labor-intensive industries. But wage structure and labor market policies must also be designed to ensure fuller employment and higher incomes in urban areas. To do this, governments must find ways to influence the pattern of rural-to-urban migration and to raise the rate of urban job creation above urban labor force expansion.

There is a growing consensus that public policies must stimulate the growth of small- and medium-scale enterprises and help informal-sector activities in urban areas. In the poorest developing countries, large-scale manufacturing is unlikely to grow fast enough to absorb the rapidly expanding urban labor force. The case for policies favoring small-scale enterprises, as Little (1987, 208) argues, rests "on evidence that small units on average use factor inputs more productively than their larger counterparts, so that a shift of resources in favor of small units would yield a net increase in output as well as an increase in the demand for unskilled labor."

The critical policy issue for governments in most developing countries in the short run is not unemployment per se but rather developing policies and programs in conjunction with the private sector to:

- Reduce the high levels of urban underemployment
- Reduce the rising levels of unemployment among better-educated young workers entering the urban labor force
- Raise national industrial productivity
- Increase the quality of and incomes derived from formal- and informal-sector jobs in urban areas

Over the longer run, governments in developing countries will face more serious challenges of stimulating private-sector investment and production to:

- Diversify urban economies in manufacturing, trade, commerce, and services and stimulate the growth of small- and medium-scale enterprises that will have to absorb the majority of new workers in cities and towns
- Provide the urban services and infrastructure needed to allow private enterprises to operate efficiently in cities and towns and to increase their output and productivity
- Allow informal-sector microenterprises (especially street traders, food preparers, hawkers and vendors, and transportation service providers) to operate more effectively and to expand their operations

In many developing countries, entrepreneurs wishing to establish or expand small- and medium-scale enterprises in urban centers experience numerous obstacles, including:

- Limited access to and the high costs of credit
- High hidden costs of start-up and operation
- Limited skills and access to information and equipment needed to improve production, efficiency, and product quality
- Difficulties of establishing and maintaining adequate marketing networks and of reducing long marketing chains
- Inability to respond quickly to changes in market demand
- Difficulties of competing with large-scale firms that benefit from preferential government policies and programs
- Limited managerial skills among small- and medium-scale business operators

Public policies and public-private sector partnership programs must help overcome these obstacles if urban job-creating enterprises are to expand.

At the same time, the newly industrializing countries will face problems over the next 35 years similar to those in Western industrial countries. Singapore and South Korea, for example, are already seeing a decline in the supply of young skilled workers because of decreasing birth rates, longer periods of education, and lower participation rates of youth in the labor force. Skilled and unskilled urban workers face more uncertain employment prospects because of changes in the structure of the economy, rising labor costs, and greater international competitiveness (Park 1988).

Research Implications

Given these complex challenges of creating sufficient numbers of jobs to absorb the rapidly growing urban labor force, further research on urban employment in developing countries must be undertaken in three areas. We need better empirical analyses of job creation requirements in developing countries, better evaluations and assessments of the impacts of population redistribution and job creation policies, and better methods of assessing the feasibility of new job creation policies in the public and private sectors.

Clearly, governments need more detailed and comprehensive empirical analyses of urban job creation requirements. Direct and surrogate indicators of changes in the economically active population and of capacity to expand employment opportunities must be developed to estimate more accurately the number, location, and types of jobs needed in urban areas. Unemployment and underemployment statistics in most countries are inadequate. Few countries have accurate estimates of job creation growth rates in major industries. Governments and private-sector firms must be able to understand the trends in production and growth in primary, secondary, and tertiary sectors and the implications for job creation in urban areas.

If governments and the private sector are to succeed in appropriately expanding urban job opportunities over the coming decades, they need better ways of evaluating the impact of population redistribution and job creation policies. Although many have experimented with policies to influence population distribution through spatially targeted development programs and expand employment opportunities through macroeconomic and investment incentive programs, these and the other policies discussed earlier have had variable results, making it difficult to assess their efficacy.

Finally, international assistance organizations, governments, and private enterprises need better methods of planning, formulating, and implementing urban job creation policies. Policymakers in developing countries need more refined methods to assess the feasibility of policy alternatives to influence population distribution and employment expansion in cities and towns.

Analysts must be able to identify more accurately the critical variables affecting labor force size and location, the direct and indirect effects of urban employment expansion policies, and interventions that can create an economic environment conducive to job creation in urban areas.

6

Small-Enterprise Promotion as an Urban Development Strategy

RAY BROMLEY

The rapid population growth of the major cities in developing countries exerts great pressure for the generation of income opportunities to enable people to satisfy their basic needs. It has long been evident that government and big business cannot provide jobs for all, and their projects often fail to reduce urban unemployment because they attract new migrants to the cities. Thus, job generation through small enterprises is crucial to minimize unemployment, relieve poverty, and raise living standards. Small enterprises build on people's creative potential, and their operation helps to foster a competitive free market for goods and services. Because they are generally low-profile, low-status, unspectacular, and labor-intensive, their proliferation and growth have little impact on the volume of in-migration to a city.

For purposes of this chapter, *small enterprises* are considered to be economic activities or businesses with fewer than 20 persons contributing their labor power and no currently valid legal claim to be part of a larger enterprise. These enterprises may be independent operators, subcontractees, or franchisees and may rely on specific larger enterprises for supplies, outlets, credit, or the rental of premises or equipment. Self-employment, partnerships, family enterprises, cooperatives, and firms based on wage labor all fall into the small-enterprise category, provided that between 1 and 19 persons use effort and skill in the enterprise in pursuit of monetary or material gain.

With the current wave of privatization, debureaucratization, and deregulation, small-enterprise promotion is quite popular in right- and left-wing circles. It contributes to economic growth and job generation and fosters competition in the supply of goods and services. Small-enterprise promotion fits the New Right of the Reagan-Thatcher-Bush era and the emerging democratic socialism in Eastern Europe and the republics of the former Soviet Union. It represents popular capitalism and widespread opportunity, projecting an ideology of hope in which with hard work and talent most people can own property, become entrepreneurs, and make profits.

All of the earliest enterprises were small, but in the modern world system small enterprises coexist and compete with larger enterprises in capitalist and socialist economies. They are strong in adaptability, innovation, and competitiveness, filling many economic niches vacated by larger enterprises and giving access to income opportunities, on-the-job training, and entrepreneurial skills. By definition, small enterprises are human in scale and at best are congenial to work in, convenient for clients, cheap, strongly competitive, and favorable to the development of entrepreneurial skills and technological innovation. New entrepreneurs often emerge as former apprentices and employees set up their own businesses, and creativity and effort levels are characteristically high. Seen as a population, however, small enterprises have high birth and death rates, with only a small minority ever growing into larger concerns.

Small-Enterprise Promotion: Removing Regulatory Constraints and Providing Supports

Despite the enormous importance of small enterprises and the variety of problems they face, there is little consensus on the policies to adopt toward them. Governments can promote small enterprises by removing regulatory constraints, for example, by simplifying bureaucratic procedures and eliminating police harassment, and also by providing supports, such as technical assistance and low-interest credit. Such promotion may directly favor small enterprises, or it may just level the playing field, giving small enterprises opportunities already provided to large ones.

There are strong economic arguments for government promotion of small enterprises. The workers and their families are usually taxpayers—perhaps through direct income or head taxes, but more importantly through value added or sales taxes on goods and services. Through their labor and consumption, these workers contribute to the national product, and often they also serve the state through the armed forces, the civil defense, or voluntary public service. Even dependents of those working in small enterprises who are currently unemployed, engaged in occupations with very low remunerations, or performing unpaid domestic labor may play a vital role as members of the reserve army of labor, available for possible future incorporation into the wage labor force and so provide downward pressure on wages in the total economy.

Some small enterprises pay taxes as firms, many pay registration and licensing fees, and those who sell to the public help to hold down the cost of living and increase their customers' quality of life. As subcontractees, franchisees, and pieceworkers and by reducing costs of living for corporate and state employees, small enterprises contribute directly to increasing the profitability and competitiveness of larger enterprises, helping them make a greater contribution to national product and exports. Finally, small enterprises contribute significantly to national human resources as training grounds for technical, commercial, and managerial skills; as pools for labor recruitment by larger firms; and as components of a competitive environment in which a few small firms can grow into much larger enterprises.

Given the strength of these arguments for government promotion of small enterprises, it may seem curious that in most countries and contexts promotion is limited and discretionary. It is important to remember, however, that small enterprises are numerous and highly heterogeneous.

The highly selective nature of most government promotion is accentuated by widely applied distinctions between manufacturing (the secondary sector) and services (the tertiary sector). Many intellectuals and officials of left- and right-wing persuasions argue that manufacturing activities are more productive and desirable than service activities. Manufacturing is considered directly productive in creating value, whereas services are felt to be unproductive or only indirectly productive, merely recirculating value. Such views are highly questionable and largely semantic in origin (see Hirst 1975, 221-230), and whether or not services can be produced like goods, some services are essential. There is a widespread tendency to view the tertiary sector as over-inflated and basically parasitic, consisting mainly of activities "bearing no observable relationship to effective labour demand, [where] the supply of labour creates its own employment opportunities" (Bhalla 1973, 288). Such negative views are quite common among many governmental officials in relation to street services (shoe shining, hawking, etc.). They are reinforced by city planning and real estate interests who want a spacious, uncongested, beautiful modern city and foster a concept of such occupations as "people who get in the way" (Cohen 1985).

The small enterprises governments most favor are artisans, small industries, and repair workshops. They are almost universally viewed as productive, labor intensive, and skill generating and many act, or can act, as subcontractees of large-scale manufacturing and distribution firms. Governments are most likely to promote small enterprises producing folkloric handicrafts because they preserve age-old skills and designs, their output cannot be mass produced, and they boost tourism.

Synthesizing the conclusions and recommendations of such authors as Farbman (1981), Harper and Tan (1979, 89-115), McGee and Yeung (1977, 113-118), Neck (1977), and the World Bank (Gordon 1978), I (Bromley 1985, 253-254) identified 15 major lines of support to small enterprises:

1. Designing and building an urban environment favorable to small enterprise—without segregating small-scale manufacturing and commerce from housing and with numerous sites for small enterprises, high-density and pedestrian areas favorable to markets and street traders, and a transport mix favorable to paratransit operators
2. Helping to create new small enterprises by identifying potential business opportunities and entrepreneurs and providing seed capital and other forms of start-up assistance
3. Promoting the organization of small enterprises into trade unions, cooperatives, associations, federations, or larger self-managed firms
4. Providing technical and managerial training and advice
5. Giving preference in allocating public services (electricity, water supply, sewerage, phone, garbage disposal, etc.) or establishing preferential tariffs
6. Providing credit at market or subsidized interest rates (most small enterprises currently pay much higher rates than those considered normal for larger enterprises with better connections and collateral)
7. Providing free or subsidized raw materials or preferential access to products handled by government marketing boards and state monopolies
8. Providing free or subsidized equipment or preferential access to scarce imported equipment
9. Providing suitable premises with subsidized rents or purchase prices
10. Providing free or subsidized advertising or an interenterprise and interinstitutional brokering service to link small enterprises to potential purchasers of their goods and services

11. Directly purchasing small-enterprise goods and services for government agencies, marketing boards, export corporations, and consumer organizations
12. Imposing constraints on competing larger enterprises and sometimes reserving specific economic activities for small enterprises
13. Prohibiting competing imports or imposing high tariff barriers on them
14. Conducting research on appropriate technologies, organizational forms, and marketing opportunities for small enterprises and diffusing the results
15. Educating the public on the merits of small enterprises and their goods and services, and providing awards and exhibition sites for outstanding ventures

Whenever small enterprises are selected or apply for support, the eventual distribution of assistance tends to favor larger and more accessible enterprises and those with more articulate and well-connected entrepreneurs. However much those who direct small-enterprise support programs may wish to achieve total coverage of enterprises, they are hampered by lack of adequate information, staff, funds, and transport and by various legal concerns. Comprehensive and current listings of small enterprises are extremely rare and hard to compile. Even when they do exist, it is difficult to ensure that the basic information announcing a support program reaches all at the same time. Further, it is precisely the smallest and most remote operators who are most likely to be illiterate or not know the language government officials speak; to lack identity papers, licenses, and other documents needed to apply for support; and to be unable to afford to stop work for a few hours to visit a government office. Very often, when some significant support is available, it has all been exhausted before those who operate the smaller, more remote enterprises have applied for it or even become aware of it.

Government support to small enterprises can be supplemented by efforts of such private and voluntary organizations (PVOs) as foundations, church groups, and worker federations. In some countries and economic activities, PVOs are the only entities providing support to small enterprises. Often PVOs have the most interesting, innovative, and effective small-enterprise programs. Their programs are often useful experimental or model schemes, but it is unrealistic to expect PVOs to meet more than a small proportion of the needs and expectations of small enterprises. Only national governments, sometimes aided by bi- or multilateral agencies, have the resources and powers to provide most small enterprises with significant support. PVOs, however, are far more likely to spread ideas and research findings internationally and to change the prevailing negative views and low priorities assigned to small enterprises.

The ILO Informal-Sector Promotion Strategy

The most representative and well-known small-enterprise promotion programs derive from the wave of policy research on the urban informal sector in developing countries inspired by the International Labor Office (ILO). The seminal policy statements appear in the ILO country mission reports on Colombia and Kenya (1970, 1972) and propose comprehensive promotion strategies for small enterprises using most of the 15 types of support policies enumerated above and also some relaxation of regulatory constraints.

From its origins in the early 1970s, the concept of *informal sector* has suffered definitional problems. Its "invention" is generally attributed to Keith Hart (1973). Hart spoke of income opportunities rather than sectors and defined informality simply as "self-employment." To Hart, informal income opportunities were ways for the poor to get by when neither corporations nor governments could provide sufficient employment for the expanding population.

Hart's basic idea was substantially enlarged by the ILO team that wrote the Kenya report (1972) and made government support for "the urban informal sector" a major element in its recommended national development strategy. The team divided the economy into two sectors, formal and informal, effectively recreating and renaming Lewis's (1954) dualistic model of the interaction between the modern and traditional sectors in economic development. They defined the informal sector using seven criteria and recommended simplified licensing procedures for informal-sector enterprises, plus a major increase in government spending to support these enterprises with technical assistance, training, and credit. In general, the ILO team saw small-scale industrial, handicraft, and repair establishments as the core of the informal sector and the priority for government support. Thus, informal-sector promotion was little more than a renaming and intellectual rationale for a policy widely advocated since the early 1950s—small-industry promotion.

Through a judicious combination of timing, clarity, publicity, and the prestige and connections of its main authors, the ILO Kenya report became a global trendsetter. Thousands of positivistic informal-sector studies were prepared in 1973-1983—some in virtually every Third

World country—with substantial funding from the ILO, the World Bank, and other international and bilateral aid organizations. Many refinements of the initial ILO multicriteria definition of the informal sector were made (e.g., Sethuraman 1976). Government support for the informal sector, particularly small-scale industry, was increasingly perceived as the key to Third World development, mobilizing grass roots entrepreneurship and harnessing intermediate technologies.

Scholarly analyses of the 1970s informal-sector literature mainly focused on whether—and how—the formal and informal sectors are related to each other. Early works implying that the informal sector is autonomous and an economy in its own right were soon shown to be misleading, and most authors opted for a dominance-dependence concept showing the two sectors to be closely interrelated and interdependent through subcontracting, franchising, and labor and capital flows. In such a model, the burden of risk and instability rests on the informal sector, and its output and employment levels depend heavily on requirements of big business (see Portes, Castells, and Benton 1989; Tokman 1978). Relatively early, however, a few authors (e.g., Breman 1976, Bromley 1978) seriously questioned the utility of the formal/informal division and the feasibility of using multicriteria definitions for the informal sector. Refinements and alternatives were suggested (e.g., Bromley and Gerry 1979; Steel 1977), but none caught on with the main body of informal-sector researchers and policymakers.

By the late 1970s, debates on the meaning and utility of the formal/informal division had developed into a dialogue of the deaf in which more and more was written but apparently very little was read. The informal-sector concept remained nebulous, and although every user applied a slightly different definition, these seemed to serve primarily as means of identifying small enterprises worthy of government support. Some advocates of the concept referred briefly to the presence of disreputable and illegal occupations within the informal sector, such as prostitution and narcotics trafficking, and many alluded to occupations of questionable social value, such as shoe shining and bar tending. In general, however, the authors passed over these occupations very quickly and without reference to public policy and then concentrated on artisans, small-scale industries, and the need for support.

The attitude of most advocates of the informal-sector concept who have followed the ILO line is that definitional questions are unimportant and the sector's existence is just as obvious as that of suburbs or a middle class. Analysts may vary in their definitions, but such differences of opinion certainly do not negate the phenomenon's existence and the concept's utility. Lisa Peattie (1987) brilliantly analyzes this line of argument and shows that such an ambiguous concept with so many definitional problems has grown in popularity because it is useful to many people and generates a coalition of interests. The ILO version of the concept has promoted a social-democratic and reformist agenda for developing countries with a slight "green" or "neoanarchist" flavoring. It has provided encouragement for indigenous enterprise, local self-help, and appropriate technologies and has advocated increased government effort and expenditure on promoting small enterprises through supports such as credit, technical assistance, and management training.

De Soto's Counterstrategy for Promoting Small Enterprises

The ILO informal-sector approach penetrated Latin America soon after the Kenya report appeared. The ILO in Geneva, its Regional Program for Latin America (PREALC), the World Bank, the Friedrich Ebert Foundation, and the International Development Research Center all supported pilot studies, and for over a decade it seemed that the multicriteria, positivist view of the informal sector prevailed in Latin America as it did in Africa and Asia. Many leading politicians and officials were still skeptical about small-enterprise promotion, but those favoring it adopted the informal-sector terminology and focused on supporting small-scale manufacturing and repair establishments.

Since the early 1980s, Peruvian economist and entrepreneur Hernando de Soto has led a growing countercurrent on small-enterprise promotion in Latin America. De Soto (1989) ignored the positivist ILO-linked literature and definitions of the informal sector and developed a concept of *informality* based on a single ends-means criterion—economic activities whose ends are licit, but whose means are illicit. Such activities generate incomes and satisfy basic needs, they involve no immoral or criminal acts like theft or murder, and yet they contravene official regulations.

To de Soto, those who practice informality are "informals"—petty entrepreneurs desperately struggling to make a living in the face of stifling governmental regulation, the horrific mismanagement of public enterprises, and manipulation of the state apparatus by economic vested-interest groups he calls *mercantilists*. He advocates deregulation, debureaucratization, and privatization as keys to economic development, suppressing mercantilist groups and opening the economy to the full force of international competitive markets. Such changes, he argues, would enable the informals to "formalize" their operations and compete more effectively, inaugurating a "free market economy" and generating rapid, self-sustaining economic growth. He recommends promoting large and small enterprises by removing regulatory constraints rather than providing supports. Thus, in sharp contrast to ILO-style informal-sector promotion, his strategies are likely to reduce total public expense.

Obviously, simplification of administrative processes and removal of absurd laws will have a positive impact, but it is far from clear that it will dramatically transform the circumstances for economic development. De Soto's principal policy recommendation is the conversion of de facto status to de jure, giving petty entrepreneurs greater security and status within the system and letting them use their property as collateral to obtain loans. This conversion would greatly increase legal property ownership, so that most adults could aspire to own real estate, a business, shares, or savings. Given the massive poverty and indebtedness of most developing countries, however, it is unclear how much large numbers of de facto to de jure shifts will achieve. In practice, most unlicensed enterprises keep on functioning, most squatter properties continue to be used, and there is active buying and selling of these properties even without official permits and titles. Some de facto property owners fear the shift to de jure status because they see it as the first step to introduction of new taxes. Inability to pay taxes, fees, and utility charges may lead them to sell up and become wage workers and tenants or to mortgage their businesses and homes, risking bankruptcy and foreclosure.

Over the last 5 years, a new version of ships passing in the night has occurred on many occasions, as researchers and policymakers using de Soto's concept of informality converse with their counterparts using variants of the ILO informal-sector concept. The conversation runs on as if both groups were talking about the same thing. In reality, the universes described are very different, and the objectives and means of small-enterprise promotion are almost totally incompatible from a purely fiscal standpoint. The cheapness of a de Soto approach to promoting small enterprises contrasts markedly with the relatively high cost of ILO-style strategies, which give less attention to removing regulatory constraints and prescribe a battery of costly supports.

Sometimes the two sides have come closer together when the ILO-style advocates reorganize their definitional criteria to emphasize the lack of appropriate documents and permits, relegating the other criteria to secondary supporting status. Pursuing such an approach, however, means that the informal sector disappears as soon as its enterprises are documented and licensed or as requirements for such procedures are abolished.

Needed Research and Economic Realities

Our ability to design effective small-enterprise promotion policies is gravely limited without appropriate national research and data sources on several important themes: (a) we need detailed studies of the social relations of production (relationships among capital, technology, and labor) for many vital but little-known types of small enterprise, notably home building, fuel and water distribution, urban public transport, groceries, and recycling; (b) several key processes—notably franchising, disguised wage work, child labor, and the gender division of labor—must be studied in greater depth and for a broader range of Third World occupations; (c) the action research field of sociolegal studies and debureaucratization that de Soto and the Institute for Liberty and Democracy in Peru developed merits far more attention in many countries; (d) new types of small enterprises and successful promotion programs must be more effectively documented and evaluated using comparable cost-effectiveness criteria; (e) definitional criteria and enumeration procedures must be improved in most countries to permit more accurate assessment of the number and significance of small enterprises, including part-time work, second occupations, seasonal work, subcontracting, and franchising; and (f) key economic transformation and experiences such as Bolivia's 1985-1987 economic stabilization must be monitored to determine the impact of macroeconomic changes on small enterprises.

If governments in developing countries minimize the regulatory constraints they put on the establishment and growth of small enterprises and amply increase their support, we can expect growth in numbers of small enterprises and in their total turnover. But we should not expect small enterprises to take over the whole economy, creating some sort of small-enterprise-based model of economic development. Small enterprises can sell and repair consumer durables like cars, televisions, and blenders, but realistically they cannot produce these goods. They can build and repair houses, but not airports, freeways, and skyscrapers. They can extend credit to some of their clients, but they cannot provide the services or security of a major bank or insurance company. In short, we will always need large firms, and small enterprises will either fill the economic niches corporate executives perceive as less profitable or complement large enterprises by taking on labor-intensive roles as subcontractees and franchisees.

Proliferation and growth of small enterprises will satisfy numerous currently unsatisfied domestic and foreign consumer demands and increase the pool of potential subcontractees and franchisees in the economy. Small enterprises pose few threats to big business, provide some creative possibilities for their expansion through vertical interlinkages, and increase the potential for a process of economic growth based on mechanisms of circular and cumulative causation. Increased economic activity in small enterprises raises the number of workers, the total income generated, and participants' purchasing power—all of which reinject money into the economy, generating more demand for the products of small and large enterprises. If small enterprises can be brought above ground, they can also contribute to tax revenues, helping to finance public services and eliminate fiscal deficits.

An optimistic scenario depicting virtually unlimited growth as a result of the promotion of small enterprises can run into three very serious problems. First, the sheer number of small enterprises and the great variety of supports that could be provided make an ILO-style multisupport strategy quite expensive if it is to achieve genuine national coverage. It would increase the size of the public sector and require the recruiting and training of large numbers of talented, honest staff. Further, it would require action research to develop new knowledge and sensitivity within government on the most appropriate operating methods for small enterprises. Many powerful vested-interest groups are likely to oppose these new expenditures, fearing either new business competition or reduced resources for their own pet projects.

Second, the strategy can collapse completely if macroeconomic or macropolitical circumstances are inappropriate. In situations of political unrest, hyperinflation, gross corruption, civil war, terrorist ascendancy, or massive incorporation into narcotics and contraband trades, policies can become virtually impossible to implement. These circumstances lead many small enterprises to collapse due to uncertainty, sabotage, extortion, or inability to respond to dramatic changes in prices, supply, and demand. Similarly, introducing severe cuts in public-sector payrolls and expenditures following many years of economic crisis may render virtually all forms of government promotion almost totally ineffective because of the demoralization of public servants and the flight of more talented and honest personnel. Currently in many developing countries, partial or total collapse of civilian government and widespread loss of respect for national authorities represent far more serious threats to economic growth than the voraciously expanding state apparatuses conservative critics describe.

Third, the strategy will probably gradually lose steam unless additional exports or alternative sources of foreign exchange can be generated. The capacity of small enterprises to satisfy currently unsatisfied demands in developing countries is limited by the low elasticity of demand for many basic goods and services and by the potential for profiteering by oligopolies handling such key raw materials as steel, cement, fuel, fertilizer, and imported grain. Further limitations are imposed by the international financial climate, which puts considerable pressure on national governments to run a balance of payments surplus and ensure an outflow of foreign exchange to service the foreign debt. To counter these problems, a significant proportion of small enterprises must earn foreign exchange by catering to external demands and expanding into such areas as crafts exports and sales to luxury tourist hotels. Most of the highest potential areas to generate foreign exchange and broaden demand will require subcontracting or franchising plans because of the scale and complexity of the markets to be supplied.

Conclusion: Building a World of Diversity and Complementarity

Small enterprises play, and certainly will continue to play, a vital role in the economics and labor markets of rapidly expanding cities and in satisfying basic needs in developing countries. Creative support policies and removal of inappropriate controls can further strengthen their role. Nevertheless, it is important to recognize that they coexist with medium-sized and large enterprises in one economy, and there are numerous and complex linkages between different types and scales of enterprise. Policymaking must be for the whole economy rather than just one sector, and policies must be adapted to the great diversity of small enterprises and links with large enterprises.

A sense of realism about prospects for small enterprises requires the recognition that high business creation rates and numerous success stories are paralleled by high failure rates and frequent collapses—the very nature of free markets and atomistic competition. Bechhofer and Elliott (1981, 185) expressed these harsh realities well:

> Survival [of small enterprises] has rested upon continuous adaptation—not always at the level of the individual petit bourgeois, for many have gone to the wall—but at the level of the stratum as a whole. Adaptation has often been born of dire necessity as migrants flock to towns only to find there is no regular work to be had, or as workers are thrown onto the dole queues in times of recession. As with all small fry, the persistence of the collectivity masks the fact of high mortality, and of depredations by governments and larger enterprises which pose continuous threats to the aggregate's survival. But so far it has survived and it is hard to accept that there are any ineluctable forces which will destroy it.

Small-enterprise failure rates can be reduced, though never eliminated, if governments create a climate of political stability and continuous economic growth, with a stable currency, low inflation, a realistic price structure—free from major subsidies, dumping, or oligopolistic speculation—and balanced rural-urban terms of trade (see Kelley and Williamson 1984, 180-185). Such conditions, characteristic of exceptional growth successes like Singapore, are relatively rare and very difficult to generate, but recognizing their importance is vital on the difficult road toward reform. The small-enterprise tail cannot wag the macroeconomic dog, and the potential contribution of small enterprises in economic growth, employment, and satisfying basic needs totally depends on the existence of an appropriate macroenvironment.

Small enterprises in Third World cities are more diverse and on average much smaller and operated by much poorer people than their equivalents in cities of advanced capitalist countries. Nevertheless, big business, enormous socioeconomic inequalities, and the growing internationalization of economic activity are ever-present realities in developing countries, making simple small-is-beautiful doctrines increasingly anachronistic and irrelevant. Many small enterprises are tied to larger companies through such practices as borrowing, renting, subcontracting, outworking, commission selling, and franchising, and an urban economy functions as a complex hierarchy of accumulation, harnessing small enterprises to the objectives of larger ones.

There is not one "reality" for small enterprises, but many. Situations are strikingly different in contrasting countries and contexts and for different types and scales of economic activity. Further, interpretations of specific situations may vary greatly depending on the ideological perspectives employed in their analysis. Small enterprises are diverse, dynamic, and controversial, and it would be absurdly idealistic to hope that somehow all the many elements can be fitted together like a jigsaw puzzle to produce a consensus on perspective. Nevertheless, there seems to be a growing agreement that supports must be increased, constraints reduced, and repressive and obstructionist policies confined to the minority of small enterprises that clearly contravene important legal and moral codes.

In promoting the desirable majority of small enterprises, numerous problems still arise in defining which forms of promotion should take priority. De Soto's advocacy of removing regulatory constraints and converting de facto to de jure status is useful but insufficient here. At the same time, the full battery of small-enterprise support policies that could emerge from ILO-type informal-sector promotion strategies poses serious problems of political viability, staffing, cost, and quality. The crucial issues are how to generate a macroeconomic and macropolitical climate favorable to the growth of small enterprises; how to ensure sustained and consistent application of government promotion policies; how to encourage simple, easily replicable pilot programs; and how to encourage the diffusion of successful innovations among small enterprises.

7

Privatization of Urban Services and Infrastructure in Developing Countries

An Assessment of Experience

DENNIS A. RONDINELLI
JOHN D. KASARDA

Developing countries throughout the world are facing increasing demands for public services, infrastructure, and shelter in urban areas. Pressures are arising from a variety of sources for central and local governments to provide a wider range and better quality of social services and physical infrastructure in cities. Exploding urban populations, the competition for economic activities among metropolitan areas, the rapid growth of secondary cities, the strains of inadequate and deteriorating physical facilities, and social pressures to expand the stock of housing all contribute to increasing demand. At the same time, governments are constrained in meeting these needs by inadequate revenues and high levels of debt service. Overburdened central ministries often provide services and infrastructure inefficiently, and many state-owned enterprises (SOEs) heavily involved in urban service delivery generate losses rather than revenues for the central treasury. In many developing countries, SOEs account for large amounts of external borrowing and absorb capital from financial markets that could be used by private investors to expand businesses that create jobs and increase tax revenues.

All these problems are leading international assistance organizations and governments in developing countries to reassess the potential of the private sector to take a more active role in financing and providing public services (e.g., health, education, and transportation) and physical infrastructure (e.g., roads; highways; utilities; telecommunications; and water, sewer, and waste treatment facilities) (World Bank 1988b).

Although privatization has received much attention over the past few years, the concept is neither new nor unique to developing countries. The private sector has delivered services and helped to build shelter and infrastructure in Western industrial nations for centuries and plays an important role in providing services and housing in many developing countries. Throughout the history of the United States, for example, private organizations have been heavily involved in building toll roads, turnpikes, canals, waterways, railroads, and airports. According to the National Academy of Public Administration (1989, viii), "Even such inherently governmental functions as tax collection, mail delivery and spying were performed by private contractors through much of the nineteenth century."

In the postcolonial period after World War II, however, governments in many developing countries nationalized private enterprises, centralized control over service provision and infrastructure investment, and took a strong role in managing their economies. But the financial problems that arose with the worldwide recession of the late 1970s, the debt crises faced by many African and Latin American countries during the early 1980s, and the shift to market economies in Eastern Europe during the early 1990s drew renewed attention to the role of the private sector in providing services and infrastructure.

In the 1980s, governments adopted privatization policies in most developing countries, efforts ranging from the limited and halfhearted attempts to divest themselves of a few money-losing state enterprises in some African countries such as Senegal and Zambia to elaborate plans for redefining government's role in the economies of such Asian countries as Malaysia and Pakistan. Indeed, with the announcement of Malaysia's sweeping plans for the sale of many SOEs, and the "corporatization" of other government agencies and public enterprises, Prime Minister Mahathir Mohamad declared, "Privatization is now the instrument of economic management of the nation" (Business Asia staff 1991, 88). The Pakistan Ministry of Finance is seeking to privatize 115 industrial companies and 4 banks by late 1992 and to reduce substantially the state's role in producing goods and providing services. Other Asian governments are experimenting with various arrangements for eliciting private-sector participation in urban shelter, social services, and physical infrastructure provision (see Rondinelli and Kasarda 1991).

Service and infrastructure needs are increasing rapidly in urban areas, but governments in most countries lack adequate financial resources to meet them. Although the private sector is already providing a great deal of shelter and other services in many developing countries, the desirability and efficacy of privatization are still hotly debated. The debate is often distorted by some advocates who make exaggerated claims that all public services can be privatized (Hanke 1987; Roth 1987). Evaluations of privatization policies indicate that many of the alleged advantages have yet to materialize (see, e.g., Van de Walle 1989; Vernon-Wortzel and Wortzel 1989; Yotopoulos 1989). Privatization has proven difficult to implement in many Asian nations, particularly those with highly centralized governments. Even in the United States and Europe, where political and economic conditions for privatization tend to be far more favorable, the managerial and financial results have been mixed (see Pack 1989).

Yet, a good deal of evidence also suggests that under proper conditions private enterprises and nongovernment organizations can—and, indeed, already do—play a crucial role in improving and expanding urban services, shelter, and infrastructure. In the future, governments in countries with high urban population growth and expanding urban economies almost certainly will have to depend more heavily on private-sector participation to meet growing needs for urban housing, services, and facilities.

Few comprehensive or systematic analyses have been made of the comparative advantages of private-sector participation in service and infrastructure provision. Here we review the experience with privatization in developing countries, focusing on Asia. We trace the reasons why governments are considering privatization, examine some of the means by which they are privatizing urban services and infrastructure, and discuss the potential advantages of and obstacles to privatization. We then identify conditions under which privatization and expanded private-sector participation in service, shelter, and infrastructure provision are appropriate and suggest ways of improving policy implementation.

Reasons for Privatization

Governments in developing countries have been experimenting with privatization for various reasons. Clearly, international lending institutions such as the World Bank and Asian Development Bank have been pressuring governments to privatize as part of overall structural adjustment reforms. Yet these pressures have also been reinforced by domestic changes, including: (a) increasing demands for urban services and shelter because of rapid urban population growth, (b) increasing financial constraints on governments in meeting public service demands, (c) inefficiencies in and financial deficits of state-owned enterprises, and (d) dissatisfaction with the quality of public services provided by central government agencies and with the results of central planning for infrastructure investment.

Demand Created by Urban Population Growth

Growing demands for urban services and infrastructure reflect dramatic increases in developing countries' urban populations, a demographic trend that will continue well into the next century. The United Nations (UNDIESA 1988b) estimates that between 1980 and the end of the 1990s, the urban population in developing countries will have doubled from a little less than 1 billion to about 2 billion. During the following 25 years, it will double again to about 4 billion. By the end of the 1990s, nearly 300 cities in developing countries will have more than 1 million residents. Demand for infrastructure and services will also increase because of the growing economic importance of productive and commercial activities located in urban areas. In much of Asia, Latin America, North Africa, and the Middle East, urban economic activities contribute more than 60 percent of gross domestic product (World Bank 1989c). Without adequate services and facilities, urban economies stagnate or decline.

Financial Constraints on Governments

A second reason for increased interest in privatization is that governments are facing tighter financial constraints on meeting the growing needs for urban services. The problems are especially difficult in countries experiencing high levels of population and economic growth in cities. For example, the Bangkok metropolitan area is growing so rapidly that demands for services and infrastructure are far outstripping public investment capacity. Observers report:

> The Electricity Generating Authority of Thailand, facing sharply diminished power reserves, is expected to spend the equivalent of $3.9 billion over the next seven years to meet power demand. Water treatment is expected to cost

> $2.9 billion over the next decade. A rapid transit system in Bangkok is expected to cost more than $1.6 billion. (Berthelsen 1989, 12)

Many of the financial difficulties developing countries now face are the result of previous nationalization policies. With nationalization and centralization of responsibilities for public service provision during the 1960s through the early 1980s, the number of public employees and the public wage bill grew rapidly. In Indonesia, for example, the number of people employed by the government increased by more than 7 percent a year from 1978 to 1986, a rate two to three times higher than the growth of all job opportunities (Research and Documentation Center for Manpower and Development 1988). Public and publicly guaranteed external debt increased rapidly. Between 1970 and 1989, public external debt in Bangladesh rose from $15 million to almost $10 billion; in India from nearly $8 billion to nearly $55 billion; and in Indonesia from $2.4 billion to nearly $41 billion. By 1989, total external debt had reached 65 percent of GNP in the Philippines, 59 percent in Indonesia, 51 percent in Malaysia, 47 percent in Pakistan, and 39 percent in Korea (World Bank 1991). Heavy borrowing and economic recession during the 1970s and early 1980s left many developing countries with little capital for investing in new infrastructure or expanding services, especially in their rapidly growing cities (World Bank 1988b).

Inefficiencies and Financial Deficits of State-Owned Enterprises

Third, privatization has become more attractive as it has become more evident that many state-owned enterprises are loss makers rather than revenue generators. World Bank studies indicate that by the start of the 1980s, SOEs in developing countries accounted for one-quarter to one-half of all outstanding domestic debt and for a substantial portion of foreign borrowing (Shirley 1983). In Thailand, 61 public enterprises accounted for over 60 percent of the government's foreign debt in 1988 (Montagu-Pollack 1990). About 40 percent of Malaysia's external debt service payments in 1987 were made by nonfinancial public enterprises (UNESCAP, Population Division 1989). The heavy demands of SOEs for capital have squeezed private investors out of capital markets in some countries and in others have limited private-sector access to borrowing for investments that could generate jobs and income in urban areas and generate public revenues for local and central governments.

State enterprises have become financial loss makers for a variety of reasons. As Vernon (1988) points out, SOEs in many developing countries encountered serious problems, including mismanagement, corruption, patronage, and padded payrolls that raised the costs of providing services and infrastructure; inefficient operations, maintenance, and service delivery, arising in part from weak competition or a monopoly position; and involvement in highly capital-intensive operations or investments with long payback periods. Many SOEs also face price constraints from governments wishing to provide subsidized or cheap services for political reasons, preventing recovery of operating and investment costs. Overly restrictive government controls on SOE budgets and finances exacerbate these problems, as do the failure of central governments to provide promised subsidies or to deliver budgetary resources in a timely manner and requirements in some countries that SOEs take over failed privately owned businesses or provide services that are inherently unprofitable.

A typical case is the State Railway of Thailand (SRT), which became a public enterprise in 1951 to fill the gap in passenger and freight transportation services. The SRT quickly became responsible for operating the country's entire rail network. With little or no competition, SRT was initially able to turn a profit. But over the years, its operating income became negative, and by the 1980s, it was the second largest loss maker of all parastatals in Thailand, exceeded only by the Bangkok Metropolitan Transit Authority (Levy and Menéndez 1990). The sources of problems clearly reflected the list of woes described earlier. Competition from other forms of transport, especially roads and highways, diminished SRT profits. SRT took on an excessive and underproductive labor force. The central government controlled tariffs and kept them below the costs of carrying passengers and freight as real costs rose. Moreover, the government required SRT to service numerous low-density routes on which it was unable to recover operating costs.

SOE inefficiencies are clearly reflected in their limited abilities to satisfy the rapidly growing need for services and infrastructure in urban areas. Government-owned telephone and telecommunications companies, for example, have been notoriously ineffective in meeting demand for services that have become crucial to cities' economic vitality. The average waiting period for telephone installation in Indonesia is nearly 8 years, in the Philippines 7 years, and in Pakistan 10 years. In Thailand customers must wait an average of 3 years to receive telephone service; by 1989, the waiting list in Bangkok alone had grown to more than 300,000 applicants. Further, even when telephones are installed, service is often poor. Call completion rates are extremely low in many developing countries because of faulty equipment and lack of switching capacity. Completion rates for trunk calls are as low as 12 percent in Pakistan and for local calls as low as 31 percent in Indonesia. Many publicly owned telecommunications companies in developing countries lack investment capital and financial resources for maintaining and expanding existing lines. Another problem is serious overstaffing. The World Bank reports that state-owned telephone companies in developing countries have 50 to 100 employees per 1,000 telephone lines in service, compared to 0.2 employees or fewer in the United States and Europe (Ambrose, Hennemeyer, and Chapon 1990).

Dissatisfaction With Centralized Planning and Government Management of Service and Infrastructure Investment

Finally, privatization is being pursued in some developing countries because of growing dissatisfaction with the quality and coverage of services central and local government agencies provide, as well as the ineffectiveness of central government planning for urban service and infrastructure investment. That dissatisfaction is reflected, for example, in assessments of the results of Indonesia's national development plans during the 1980s, which prescribed standard urban investment priorities for all its cities. Government officials now admit:

> This sectoral and centralized approach did not meet local needs in a balanced way. Emphasis was given to water supply even where, for example, drainage or solid waste management required more urgent attention. The consequent imbalance in development led later to problems with the financing of the operation and maintenance of unnecessary and unwanted services and to difficulties in financing further, more important developments. (Padmopranoto 1987, 445)

In South Korea, central government investments in infrastructure strongly influenced the urbanization pattern. Analysts argue, however, that public corporations "lack sensitivity to local needs and priorities since they operate on standard procedures formulated for the whole country" (Choi 1987, 512). Special laws and national planning criteria and procedures constrain their investment decisions. They rarely consider each city's unique needs.

Local government agencies have rarely been more effective than central ministries in providing basic urban services and infrastructure. Studies of towns in the region surrounding New Delhi, for example, show that coverage for water, sewer, and other basic urban services is extremely low. The studies indicate that government agencies can recover as little as one-fourth of the costs of providing and distributing water in many urban centers—30-40 percent of the revenues are lost because of leakage, theft, and wastage. Cost-recovery rates for public transportation services in many Indian cities are as low as 40 percent (Mathur 1989). Mathur concluded from his review of municipal services studies in India that local public agencies have generally failed to fulfill their service functions: "The services they provide are both inadequate and inequitable; there is a sizable element of subsidy in urban service provision; the cost recovery rates are low; and the administrative capacity and willingness to raise resources is woefully absent" (p. 13).

Dissatisfaction also arises from the tendency of some governments to assign a higher priority to new construction than to maintaining existing facilities. Other problems stem from citizens' perception that central government services and infrastructure are "free" goods to which they are entitled, rather than local resources they should provide for themselves or pay for. Scarce central government resources are drained from other uses to support services that could be provided and paid for locally.

Means of Privatizing Public Services

For all of these reasons, in providing services and infrastructure, central and local government roles have come under closer scrutiny. Various forms of privatization are being seriously considered as alternatives to the expansion of government responsibilities for service provision.

Privatization Defined

The concept of privatization covers a wide range of policies. Defined narrowly, it is simply the sale of public assets or enterprises to private investors. This form of privatization is being pursued by countries as diverse as Algeria, Tanzania, Malaysia, Thailand, South Korea, and

Malawi (see Nellis and Kikeri 1989). But outright sale of government corporations or the assets of public agencies is only one, and sometimes a very difficult, form of privatization. *A broader concept of privatization encompasses a wide range of policies to encourage private-sector participation in public service provision and that eliminate or modify the monopoly status of public enterprises* (see Rondinelli, Nellis, and Cheema 1983; Figure 7.1).

Using this broader concept of privatization, governments are deregulating many sectors to allow private competition with public agencies and enterprises. They are "corporatizing" and "marketizing" public agencies to require them to cover their costs and manage their operations more efficiently. They are allowing or encouraging businesses, community groups, cooperatives, private voluntary associations, small enterprises, and other nongovernmental organizations to offer social services. They are providing financial assistance and guarantees for private investment in infrastructure and facilities, creating public-private partnerships, and contracting with private organizations. These policies assume that many goods and services for which people can pay—such as public transportation, electric power, piped water, or housing—can be delivered more efficiently and effectively by private enterprises directly or through contracting than by either central or local governments (Roth 1987).

Forms of Privatization in Developing Countries

Seven forms of privatization are being used extensively in developing countries. Each has different characteristics and involves different combinations of public and private participation in providing urban services and infrastructure.

1. Divestment of State-Owned Assets. The most extreme form of privatization is government divestment of state-owned enterprises or assets. In many countries, the government simply sells all or part of its ownership in state enterprises or liquidates unprofitable and ineffective public agencies. Several instruments are used for the divesting of state enterprises, including (a) public offering of shares on the stock exchange; (b) private sale of shares to domestic or foreign investors; (c) management or employee buy-outs; and (d) distribution of vouchers or shares to all citizens or particular groups, such as company employees or low-income families (for a detailed assessment of these alternatives, see Vuylsteke 1988).

Implication of Privatization

High —— Degree of Government Divestment —— Low
High —— Potential Opposition by Government Employees —— Low
Low —— Degree of Interaction Between Public and Private Sectors in Service Delivery —— High

Forms of Privatization

Sale of public assets to private investors	State subsidies for private service delivery	Incentives or guarantees for private-sector participation	Government encouragement & support for private-sector expansion in service industry
Deregulation & liberalization policies to let private sector participate in service provision	Contracting with private sector • preformance contracts • "afterimage" system	Public-private partnerships • joint ventures • joint investments • turnkey projects	Transfer of public service responsibilities to non-government organizations (professional groups, technical bodies, cooperatives)

Figure 7.1. Forms and implications of privatization

Sale of SOEs and the assets of some public agencies is an important instrument of privatization in Bangladesh, Pakistan, Sri Lanka, and Malaysia. By the early 1980s, the Bangladesh government had targeted more than 70 SOEs for divestment (Berg and Shirley 1987). Nine SOEs were partially or completely sold in Thailand by the end of the 1980s. In Sri Lanka, 7 state companies were sold in 1990 with 22 more put up for sale in 1991. In Malaysia, the government has partially divested itself of commercial enterprises in Port Klang and of the national shipping line and national telecommunications company as well. It plans to privatize four major highways, the Malaysia-Singapore Second Crossing, and the National Institute of Cardiology. In South Korea, the government is offering up to 75 percent of the shares in 11 of the most profitable state-owned enterprises to low-income groups at below market prices. All future privatizations will require some percentage of the shares to be sold to employees of the company through Employee Share Ownership Plans (ESOPs) (Montagu-Pollack 1990).

2. *Marketization or Demonopolization of SOEs.* In some countries, SOEs providing urban services are being restructured to make them more efficient and effective and to make them compete with or operate like private companies. A range of alternatives is being tried, including

(a) eliminating subsidies and forcing SOEs to recover costs or to make a profit; (b) creating joint stock companies in which public and private investors hold shares and give direction to the enterprise; (c) allowing private firms to compete with SOEs in providing goods and services, thereby eliminating their monopoly position; and (d) breaking up SOEs into divisions, some of which are divested although others are retained as public enterprises (Vuylsteke 1988).

In Malaysia, the government has already corporatized several ports, the National Electricity Board, and the Sabah Gas Industries. There are plans to restructure the Selangor Water Supply Company, several airports, the postal service, and the railway system. Restructuring is taking place as well in Thailand, Pakistan, Singapore, and Indonesia. In the People's Republic of China, the central government has introduced measures to make SOEs more efficient: The "production responsibility system" has been introduced in more than 90 percent of the SOEs to provide firm-level managers with more discretion in production planning, pricing, and use of retained profits. Performance targets have been established for many SOEs, and proposals are being considered to enhance competition among public enterprises controlled by central, provincial, and municipal governments by requiring ownership to be spread among different public agencies and levels of government (see Lee and Nellis 1990).

The Republic of Korea began a successful comprehensive public enterprise reform program in 1983 targeting 26 companies—known as the Government Invested Enterprises, or GIEs—that accounted for 60 percent of the value added in the public sector (World Bank. Country Economics Department 1991). The reforms included relaxing government controls on the management of these enterprises and allowing managers much more discretion to make decisions on personnel matters, budgets, and procurement. Political patronage appointments were drastically reduced. All enterprises were subjected to performance reviews used to make decisions on promotions and distribution of annual incentive bonuses. Outside experts and enterprise managers set performance indicators and targets, and enterprises are ranked yearly by their performance. Performance rankings are published in the newspapers, giving managers strong incentives to adopt efficient management practices. The reforms brought dramatic improvements in SOE operations. Since the reforms were initiated, the ratio of costs of sales to total revenues declined steadily, profitability of many of the enterprises increased, and none of the enterprises incurred losses thereafter.

3. Contracting With Private Companies to Provide Public Services. In nearly all developing countries, central agencies or local governments contract with private organizations to help provide services that public agencies cannot offer efficiently or effectively. Contracting for services lets governments arrange with private organizations to provide services, facilities, or infrastructure that meet government specifications (Ferris and Graddy 1986; Savas 1982). In Nepal, for example, many town governments use private contractors to collect local taxes (McCullough and Steubner 1985). In Sri Lanka, municipal markets are rented under tender to private merchants (see McCullough 1984, 1985; Minis and Johnson 1982).

Contracting for urban services is also being tried in Malaysia. The municipal council of Petaling Jaya, a city of 300,000 people, turned to privatization when it experienced declining revenues, mismanagement, and rising costs in the collection of parking fees. In 1984, the council leased parking areas to private management firms in return for a monthly rental, thus retaining control over parking services while relieving itself of management and financial responsibilities (Tan 1987).

The State Railway Authority of Thailand (SRT) has successfully experimented with contracts with private firms to provide service on three intercity rail routes that were generating substantial losses. The private companies lease passenger railcars and railway lines from SRT and pay it a fee every 15 days. The contractors cover the costs of railcar maintenance and cleaning and optional concession services. SRT provides the use of railway stations and personnel to manage them, as well as train drivers and guards.

4. Public-Private Partnerships. In situations where the outright selling of SOEs or contracting for services is not suitable, governments are experimenting with public-private partnerships. These partnerships vary in their characteristics, including (a) joint ventures, in which public and private organizations formally or informally work together to implement urban development activities; (b) build-operate-transfer, or turnkey, projects, in which governments agree to buy or lease completed facilities constructed by private investors or vice versa; and (c) joint investment, in which both public and private organizations take an active role in financing facilities, infrastructure, low-cost housing, or urban development projects.

Privatization policies in many developing countries either require or allow the government to retain some share of the stock in profitable or politically strategic companies, making them, in effect, joint ventures.

Because of recent weak capital markets, the government of Taiwan has been selling shares in SOEs gradually, retaining majority ownership until demand for stocks increases. In Malaysia, the government continues to hold shares in nearly all of the companies that have been privatized (Montagu-Pollack 1990).

Other forms of public-private partnership are also being tried. The Indonesian Postal and Telecommunications Ministry has build-operate-transfer agreements with five private companies to install 100,000 telephone lines in metropolitan Jakarta. The companies will install and manage the lines for 9 years, retaining 70 percent of the profits to recoup their investment. They will share 30 percent of the profits with the government and at the end of the agreement turn the system over to the ministry (Ambrose, Hennemeyer, and Chapon 1990).

Hong Kong formed a public-private partnership arrangement with private firms in 1988 to redevelop slum areas and physically deteriorating parts of the city. Its Land Development Corporation (LDC) selects suitable areas for redevelopment, prepares redevelopment strategies, and formulates plans for approval of the Town Planning Board. After a plan is approved, the LDC acquires land (with fair compensation) within the area and elicits private company bids to redevelop it with a suitable combination of residential, commercial, and other uses. The process combines the advantages of public planning and urban land acquisition in a way that ensures adequate housing for displaced residents and maintains environmental standards via the private sector's superior ability to mobilize capital for investment and efficiently construct infrastructure and residential and commercial buildings (Yeh 1990).

5. Public-Private Cooperation in the Development of Facilities or Infrastructure. Less formal cooperative arrangements have also been made between governments and the private sector to meet growing demands for urban services, infrastructure, and shelter. For instance, many cities use sites-and-services programs to provide housing and basic facilities for the poor. Beginning in the 1970s, many governments sought alternatives to public housing to meet growing needs for low-cost shelter. Through sites-and-services programs, government housing agencies assembled, cleared, and provided services and infrastructure to land that was subdivided into home sites and sold to the poor. Poor families contracted with private builders to construct their dwellings or built houses themselves, usually with subsidized materials or low-interest credit (Rondinelli 1990a).

In India, federal and state government agencies are encouraging private companies to become more heavily involved in land development and low-cost housing construction. In Ahmadabad, the Parshwanath Group, a private construction and housing finance company, plays an active role in providing low-cost housing with support from local regulatory authorities. This private corporation assembles land for housing projects, obtains approvals from the Ahmadabad Urban Development Authority, helps organize cooperative societies that hold title to land and perform maintenance functions after the project is completed, and obtains mortgage financing for beneficiaries from the public Housing and Urban Development Corporation. With government assistance and encouragement, the company has been able to construct more than 17,000 low-cost housing units in and around the city of Ahmadabad (PADCO 1991).

The Hong Kong government is using such incentives as relaxing building regulations, changing land lease renewal arrangements, and modifying subdivision and zoning regulations to stimulate private-sector investment in high-rise apartments to ease the housing shortage and relocate slum dwellers. These changes seek to increase the profitability of urban redevelopment for private companies (Yeh 1990).

6. Transferring or Leaving Service Provision to Nongovernment Organizations. Some governments are transferring service delivery functions to nongovernment organizations or simply leaving the provision of some types of urban development to private enterprise. In Asia, cooperative organizations, trade unions, women's and youth clubs, and religious groups are all involved in some aspects of urban service provision (Ralston, Anderson, and Colson 1981). Nongovernment and religious organizations provide health, education, and training programs that supplement those offered by government (Rondinelli, Middleton, and Verspoor 1990). In the Philippines, for example, the Catholic church has played an important role in supplementing the public education system by operating elementary and secondary schools, as well as colleges and universities. Other religious groups run hospitals and health clinics and provide other social services that are either not available from the government or are inadequate. In Vietnam, individual physicians or groups of medical personnel have been allowed to open private health clinics, especially in crowded urban neighborhoods, to improve access to health services and relieve pressures on state hospitals that lack sufficient beds, equipment, and medicines to provide adequate health care (Hiebert 1991).

In India, the government registers and aids housing cooperatives that buy land and procure financing for low-cost housing construction.

Housing cooperatives may account for a quarter of all private formal housing production in urban areas in India (PADCO 1991, 14-16).

In the Philippines, development of one of the most modern and efficient sections of Metropolitan Manila, the Makati district, has largely been left to a private company (Santiago 1989). The Ayala Corporation, which owned the land on which the Makati Commercial Center was developed, formulated its own long-range plan for subdividing large tracts of land for residential, commercial, institutional, and recreational uses. This part of Manila contains moderate-income and exclusive residential areas. The company has developed office, commercial, and manufacturing facilities and donated land for schools, parks, playgrounds, and municipal buildings. It takes an active role in making community improvements and provides its own engineers to assist the municipal council in maintaining Makati's public infrastructure.

7. Guarantees or Incentives for Private-Sector Service and Shelter Provision. Government agencies in developing countries also offer guarantees or fiscal incentives to induce private organizations to provide services that contribute to urban development, usually by providing loans or subsidies to individuals or groups to purchase services or housing from the private sector. In India, the National Housing Bank has provided equity capital to some private housing finance companies that offer mortgages to individuals buying privately built houses. India's Housing and Urban Development Corporation has used government funds to re-lend to a private organization providing low-cost shelter (PADCO 1991, 16-18).

Although many countries are actively experimenting with these and other forms of privatization, the private sector's proper role in providing what have traditionally been considered public services is still hotly debated. More systematic evaluations of experience with various forms of privatization in developing countries are needed to guide policymaking and implementation.

Advantages of Privatization

Experience thus far with privatization in developing countries indicates that private corporations, nongovernment organizations, and even informal-sector enterprises have potential advantages over government agencies in providing certain services. These advantages include the ability of private companies operating in competitive markets to offer lower production and delivery costs; greater efficiency in service delivery; better access to current technology; greater capacity to obtain and maintain capital equipment; wider choice and more flexibility in service provision; more rapid and efficient decision making; reduced financial burdens on government for wages, operating costs, debt servicing, and investment; fewer restrictions in work and hiring practices; and more flexibility in adjusting the types and levels of services to changing needs (Hainsworth 1990, 124).

Studies in developing countries show that private companies often provide more efficient services and infrastructure in urban areas at lower costs. Privatization of parking services in Petaling Jaya in Malaysia resulted in substantially more revenues than those collected by the municipality. Average daily collections of attendants rose, and management was generally more efficient. Moreover, the municipal council was assured a predictable level of revenues from leases without the costs and the financial risks involved in managing the parking services. At the same time, efficient contractors made an acceptable profit, demonstrated by the large numbers of bids submitted to the council each time leases were auctioned (Tan 1987).

World Bank studies (1986) point out that in many cities small private bus companies and minibus operators often provide better, cheaper transport than do large municipal bus systems. Public systems do not generally offer as convenient or flexible a service in slum and squatter settlements or maintain transport routes within easy walking distance of low-income neighborhoods. In Calcutta and Bangkok, private minibuses offer more convenient and flexible service in and around poor neighborhoods, thereby attracting low-income passengers even when fares are slightly higher than those of public transport. Bangkok's minibuses are profitable although Bangkok Metropolitan Transit Authority buses incur losses. The profitability of private minibuses is attributed to more flexibility in routing and scheduling, higher labor productivity, and the ability to offer more convenient service (UNESCAP 1989). In Calcutta, private operators run buses of the same size and at the same fares as those of the Calcutta State Transport Corporation (CSTC) and make a profit whereas the CSTC needs government subsidy; the private companies' buses operate at half the CSTC's cost. Moreover, the private companies have much higher levels of labor productivity and drivers receive a percentage of revenues to make them more vigilant against fare evasion (World Bank 1986).

Experience with the SRT contracts with three private companies to run intercity rail lines in Thailand showed that the private sector could make profits on the same lines on which the SRT lost money. Contractors could raise fares and still attract new customers because they offered more frequent express service without local stops, a relatively flat price structure to attract long-distance passengers, air-conditioned cars, and meal options. The three firms targeted a segmented market of higher-income, long-distance travelers. Contracting not only generated fees for SRT, but the agency also learned from the private companies how to improve some of the operations it retained (Levy and Menéndez 1990).

Another activity that the private sector—especially small-scale enterprises and private voluntary organizations—seems to carry out more effectively and efficiently than public agencies is housing construction. Despite the large public housing programs of many governments, most of the shelter built in developing countries is constructed by small informal-sector enterprises and individual builders (Rondinelli 1986). Virtually all of the dwellings of the poor in Nepal are built by the informal sector, which also constructs a large percentage of houses in urban areas of Indonesia (Hardoy and Satterthwaite 1981). In India, the private company that provides low-income housing in Ahmadabad makes a profit by efficiently integrating its supply systems, constructing projects quickly to minimize the adverse impact of price increases for building materials, and investing efficiently in large tracts of land on the periphery of the city (PADCO 1991).

Even public officials admit that the private sector can make investment decisions more effectively than public bureaucracies. "The private sector enjoys a distinct advantage over the public sector in undertaking urban development activities," one Filipino official pointed out; unlike government, the private sector "is not constrained by rigorous administrative procedures and requirements, particularly in terms of budgeting, disbursement, accounting and auditing requirements" (Ramos 1987, 711). Private-sector decisions are less hampered by the bureaucratic controls and elaborate interagency referrals and approvals required in the public sector, so "private-sector projects get easily implemented within program targets and schedules."

Experience also suggests that nongovernment organizations and professionals in private practice can play a crucial role in supplementing public health services in urban areas. The decision to allow private clinics to operate in Vietnam has provided much greater access to basic health services than state hospitals could offer. As a result of increased private participation in health care and the introduction of higher charges in state hospitals, hospital bed occupancy rates dropped from

100 percent to 60-80 percent within 2 years, relieving some of the relentless pressures and financial burdens on government health agencies (see Hiebert 1991, 16-17).

Financial pressures on the government of Malaysia also declined as a result of privatization and corporatization. By selling shares of SOEs, the government was able to transfer nearly 20 percent of its outstanding debt to the private sector by 1987, saving more than M$3.9 billion by privatizing public works projects (Tsuruoka 1990).

Obstacles to Privatization

Despite evidence that private participation in providing urban services can reduce costs, increase efficiency and coverage, and reduce the financial and management burdens on government, opposition remains strong in many developing countries, impeding the implementation of privatization policies. Even in countries where there is strong rhetorical support for privatization, the process has moved more slowly than anticipated. Some countries established complex and ambiguous organizational structures and procedures. As a result, early attempts to assess the value of state enterprises took a great deal of time in Malaysia and the Philippines. In Malaysia, privatization has also been impeded by administrative and legal constraints. Some departments and agencies operated under their own legislation, which had to be amended through a long, complicated process to authorize them to sell off all or part of their assets. Complications arose as well from all public employees having their pension rights constitutionally guaranteed. Alternative ways of securing their pensions had to be found when companies were privatized (al-Haj and Yusof 1985). Further, the Malaysian government has used its control over public enterprises to try to eradicate poverty and promote economic opportunities for indigenous Malays, and these political objectives had to be considered with others in setting guidelines for the sale of SOEs.

In Malaysia, Indonesia, and Taiwan, weak stock markets and low demand for shares slowed the process of selling state-owned companies in 1990 and 1991. Weak trading on the Jakarta Stock Exchange and the Indonesian government's tight money policies temporarily halted the privatization of several state enterprises in 1991 (Bolderson 1991).

A far more serious problem has been that policies for privatization and for expanding the private-sector participation in urban service provision are often strongly opposed in developing countries. Opposition often comes from political leaders who control patronage positions, from civil service organizations and labor unions who fear loss of government jobs, and from some organized consumer groups who fear large increases in costs of subsidized services. In some countries, powerful business interests that are unwilling to see long-standing relationships with the government disrupted also oppose more open competition in the public service industry (Van de Walle 1989).

Entrenched political opposition to privatization remains strong in many developing countries. Members of civil service organizations and labor unions in Thailand, for example, have been outspoken in their opposition to privatization, fearing that transfer of services to the private sector will not only result in job losses but will also reduce their power in the service industry. Union leaders contend that private firms will simply substitute jobs with inferior wages and working conditions in trying to cut costs or maximize profits. To the extent that the private sector can prevent unionization, workers worry that they will have little redress for grievances and no power to influence working conditions. "Nine of Thailand's 10 largest labor unions are in the state sector. Organized labor has been largely frozen out of the few small companies that have been privatized over the past decade. That has helped to put opposition to privatization at the top of labor's agenda" (Berthelsen, 1989, p. 12).

Some of the strongest opposition to privatization in Thailand has come not from labor unions, but from the military, which uses its control over public corporations to ensure employment for high-ranking officers after they retire, and from central government ministry officials. Despite strong support for privatization by top political leaders in Thailand, various ministries and agencies have opposed, delayed, or undermined the program because they fear their agencies will lose power, status, or control over budgetary resources. At the peak of debates over privatizing ports and utilities in 1990, the minister of Transport and Communications publicly announced that he would fight privatization of any enterprises under his ministry's supervision, including the country's most important SOEs such as the Port Authority, state railway (SRT), and the Telephone Authority (Stier 1990).

Social, ethnic, and religious groups benefiting from government control over economic activities have also voiced opposition to privatization. In Indonesia, privatization is strongly opposed by "economic nationalists and the Muslim fundamentalists who abhor Western consumerism and capitalism, and those who expect government to intervene for redistributive purposes (whether in the form of food subsidies, price controls, job creation, social programs or regional development projects)" (Hainsworth 1990, 124). In Indonesia and Malaysia, there is a general fear that privatization will give greater control of service industries to wealthy Chinese industrialists and financiers or to multinational firms.

Opposition to privatization also arises from fears that private firms will eliminate services that are unprofitable, provide inferior service in trying to maximize profits, and leave poorer parts of cities or large numbers of poor households unserved. Critics point out that although private minibuses in Bangkok make a profit, they do so by dropping low-density routes and leaving the undesirable and unprofitable ones to the financially strapped Bangkok Metropolitan Transit Authority. Similarly, the private companies that leased the three intercity rail lines from Thailand's state railway made a profit by dropping unprofitable local services and appealing to higher-income customers. Moreover, opponents claim that privatization can reduce public control over the types and quality of services available. They contend that Bangkok's minibuses are more profitable because many ignore traffic regulations, overcrowd passengers, and fail to maintain equipment to safe standards (DeHoog 1984; Hatry 1983).

Although housing construction is one sector that the World Bank and almost all other international organizations have urged governments to leave largely to the private sector, critics of privatization emphasize that in many developing countries private construction companies can provide only a small fraction of the total number of housing units needed. Moreover, they argue that poor households cannot afford housing built by formal-sector companies.

Experience suggests that the success of privatization does depend heavily on the management skills of private companies (Vernon-Wortzel and Wortzel 1989), which in many countries remain weak. In Singapore, where state-owned enterprises dominated service provision for a long time, the state siphoned off the best managers for public enterprises and provided few incentives for entrepreneurs to develop experience in providing quasi-public services (Low 1988). Even countries with a strong private sector have experienced problems with privatization. The private contractors who operated parking facilities in Petaling Jaya in Malaysia were not equally successful. Some had as much difficulty as the municipal council in collecting parking fees. Some small-scale contractors could not supervise their employees adequately to ensure that all collections were turned in to the company. Some companies had difficulty paying monthly rentals to the municipal council on time (Tan 1987).

Privatization has also proceeded more slowly than anticipated because of continuing public suspicion of the motives and practices of private firms engaged in urban development activities. In India, observers point out that "The degree of private sector involvement in urban development is constrained by the perceptions that the public sector needs to operate in the interest of controlling speculation and the equitable distribution of economic opportunities" (Sukthankar and Sundaram 1987, 410). Others claim that in the past, private companies have acted irresponsibly in constructing housing projects illegally on land reserved for greenbelts or public facilities.

Opposition also arises among political leaders and consumer groups who fear the increased political power of private organizations. Critics argue that when governments privatize "natural monopolies" such as gas, electric, or other utilities, they may merely transfer monopolistic practices from the public to the private sector. This allows big businesses to obtain greater economic and political influence that cannot be controlled easily by citizens or public officials. Critics of privatization note there is little incentive for private monopolies to operate efficiently, cut costs, or act in the public interest and that they can be as unresponsive to consumer needs and political mandates as state-owned corporations.

Finally, opposition has arisen among groups who fear that privatization will let governments shift responsibility for social services to the private sector. They fear that privatization will simply be a means through which governments divest themselves of obligations for performing important but expensive or politically undesirable social functions. By privatizing, they can diffuse political pressures to deal with critical social needs. In Singapore, for instance, where government enterprises have provided almost all urban services and infrastructure, the ruling party's approach delegates responsibility for social programs such as housing and health care to private employers and workers (Lim 1983, 764). In such a situation, critics argue, privatization may get the government "off the hook," at least temporarily, for providing services and facilities or for solving urban problems that are, in the long run, beyond the capacity of private organizations.

Conditions Under Which Privatization Can Be Effective

Despite formidable obstacles to privatization in some countries, the experiences cited here suggest that private enterprises and nongovernment organizations are playing increasingly important roles in providing urban services and shelter. At the same time, it should be apparent that although the private sector offers advantages in financing, operating, and managing some urban services and facilities, it can neither simply displace government nor profitably or efficiently provide all of the services now offered by public agencies. Privatization tends to be more successful where markets are already operating efficiently or could be made to do so with effective policy reforms. It is more likely to succeed where state enterprises compete with established private businesses (Bishop and Kay 1989). The process is likely to be slow in countries lacking strong capital markets, however, where the private sector cannot easily obtain access to credit, where there is strong political opposition by government officials and public service unions, and where government limits the selection of eligible buyers of state enterprises (Cowan 1987).

Because of these complexities, government agencies and private businesses need guidelines for assessing opportunities and comparative advantages in urban service provision. We offer a framework for identifying opportunities for privatization, based on the nature and characteristics of urban services (Figure 7.2).

Limits on Privatization and Private-Sector Participation

The private sector will likely play a more limited role in, and governments will most likely retain responsibility for, those services and infrastructure that society considers collective goods. Such services include those that meet basic social needs; that political leaders feel an obligation to maintain or expand regardless of the economic costs; that the public considers essential to the health, safety, and welfare of the community as whole; and that require huge "lumpy" investments (such as highways or mass transit systems), the capital for which must be raised through bond issues or public borrowing.

Government agencies are also likely to maintain control over services that are difficult politically or economically for government or the private sector to charge users for (and are therefore not likely to be profitable to private companies) and whose delivery have potential spillover effects from one local jurisdiction to another—such as pollution and disease control.

Characteristic of Service	Service Provider		
	Government	Public-private partnership	Private enterprise
Type of service	Public goods	Public goods for which user charges can be levied	Private goods or public services for which costs can be recovered
Primary beneficiaries	Community	Identifiable groups	Individual or household
Public perception of necessity of services	Essential basic needs, merit goods	Essential services	Discretionary
Cost characteristics	Indivisible	Divisible	Divisible
Relationship between demand and willingness to pay	Low	Moderate	High
Measurability of quantity and quality of service provided	Low	High	High
"Spillover effects" of service	High	High	Low
Capital-investment requirements	Large, "lumpy"	Moderate or large incremental	Low or moderate
Capacity of nongovernmental organizations to provide services	Low	High in specialized areas	High
Technical or technological sophistication required	Low	Moderate or high	High

Figure 7.2. Framework for Assessing Opportunities for Privatizing Urban Services and Infrastructure Provision

SOURCE: Rondinelli and Kasarda 1991

As we noted previously, privatization is unlikely to occur rapidly in countries where the private sector is weak, regardless of the level of inefficiency and ineffectiveness in public agencies.

Opportunities for Private-Sector Participation

In developing countries where there *is* a strong private sector, private enterprises can and do play important roles in providing services and facilities that are, in reality, private or noncollective goods. The greatest opportunities for privatization are in services such as transportation, water provision, health care, and trash collection, where users can be identified and some groups can be excluded if they fail to pay for the services. The private sector is likely to play a larger role in providing services with weak territorial spillover effects that consumers are able and willing to pay for or that have a higher level of quality. The quantity and quality of the services delivered must, of course, be measurable.

The private sector can also find an important role in providing services for which a high level of technical or technological sophistication is required, one not easily or efficiently attained in public agencies such as utilities and energy generation.

Private organizations can find opportunities to provide even collective goods at a higher level of quality than governments can offer. For example, in some developing countries the private sector provides specialized health care in urban areas much more effectively than public institutions.

Finally, private enterprises, especially those in the informal sector, can play a large "gap-filling" role; they can and do provide services for poor households that are unserved by local governments. The informal sector usually extends services or goods that government has not made widely available, that can be offered in small lots to cater to the needs of the poor, or that are not extended to new areas of settlement. Water vendors, for example, are common in cities where municipal water systems have not been extended to slum communities. In addition, traditional healers, herbalists, and midwives provide health services where public health programs are lacking.

Nongovernmental organizations can also help in extending public services through self-help or public-private partnership programs. Cities provide ample scope for public and private organizations to work together to provide low-cost shelter more effectively through contracting and turnkey projects.

Conclusions

Governments can do much to stimulate and support private-sector participation in the provision of urban services, not only of large companies but also of small-scale and informal-sector enterprises and nongovernment organizations. The experience in developing countries thus far with privatization, contracting, and public-private partnerships indicates that to improve the privatization process some or all of the following actions must be taken:

1. Governments must remove the unnecessary restrictions and regulations that inhibit private companies and informal-sector enterprises from operating effectively in urban areas and provide or encourage low-cost credit programs that allow small enterprises to expand (see de Soto 1989).
2. Governments must offer opportunities for private enterprises to develop their management capabilities by contracting for services that are not effectively provided by public agencies. Contracting can be most effective in situations where efficiencies in service provision arise from economies of scale, where greater productivity can be obtained from hiring skilled labor or specialized professionals, and where contractors can use part-time labor or less labor-intensive methods of operation. Contracting can be effective in cities where private companies are free from the severe labor and wage restrictions that characterize many civil service systems and have the flexibility to seek minimum cost approaches to service delivery (see Ferris and Graddy 1986).
3. To promote privatization, governments may have to provide, at least initially, incentives and assurances to protect current civil service employees. Options include allowing employees of public corporations or agencies to form their own companies to contract for service provision or to acquire public assets, establishing employee share ownership plans, or requiring private companies that take over state enterprises to give preference in rehiring laid-off workers to current employees of those enterprises. Phased-in privatization by geographical district or by stages could reduce opposition. The hostility of public employee groups may be contained by ensuring that fair wages and working conditions are provided by private companies that take over from public agencies or win contracts to provide public services.
4. Governments can also allay fears that the poorest urban households will be excluded from privatized services by offering companies assistance or subsidies for providing unprofitable services for the poor. Regulated utility or transportation companies, for example, should be allowed to charge high enough rates to provide "cross subsidies" to their poorest customers. These partial subsidies may be far more economical for governments than continuing to underwrite the costs of inefficient and ineffective public enterprises or agencies. Voucher schemes that give low-income groups the ability to purchase services from the private sector directly are another means of overcoming exclusion problems and of strengthening the market for social services.

5. In some countries, the government will have to take serious measures to reform state-owned enterprises before private investors are willing to purchase shares or take over their operations. In Indonesia, for example, the Minister of Finance found in 1990 that about two-thirds of the country's 189 state enterprises were "unhealthy"—either insolvent, unprofitable, or lacking adequate liquidity (Hainsworth 1990, 124). In other countries, SOEs may have to be restructured to make them more efficient as public agencies before private companies will be interested in joint ventures, public-private partnerships, or contracting.
6. Legal reforms must precede privatization in countries where laws or regulations have been intolerant of or hostile to the private sector. Changes must often be made in macroeconomic policies and labor laws, protectionist trade policies, restrictions on access to credit, wage and price controls, and the property rights system before private enterprise is willing or able to provide services effectively.
7. Improvements in the implementation of privatization policy require a redefinition of the roles of government and public employees. A National (U.S.) Academy of Public Administration (1989, 8) panel concluded that a fundamental distinction must be made between "government as a *financier, authorizer* and *overseer* of services, and government as a producer or provider of services." The panel emphasized that "the broad definition extends the term privatization to the wide set of arrangements under which government remains involved as the financier or authorizer of services but relies on the private sector or the market for the actual provision and delivery."

Redefining the role and responsibilities of government implies a reorientation of its activities and procedures. When government acts as a *facilitator of service provision* rather than as the primary provider, it must shift from a control-oriented to an adaptive approach to administration (see Rondinelli 1992). In a market-oriented system of service provision, government cannot operate effectively using a hierarchical command-oriented system of management; effectiveness depends more on negotiation, persuasion, participatory decision making, and coordination. Government officials must develop the capacity to manage contractual relationships. Government must take on more responsibility for enforcing regulations that protect the collective welfare, ensure open competition, and promote the advantages of market discipline without strangling the market with unnecessary or unrealistic controls. Public officials and employees must be trained in adaptive management, negotiation and interaction, effective regulation, and in understanding how private companies operate.

Experience suggests that although privatization is not a panacea, under appropriate conditions the private sector can play a cost-effective, valuable role in meeting growing urban service and infrastructure needs. The challenge facing government and business leaders in developing countries over the next decade is to find the proper balance between public- and private-sector responsibilities for providing the services, shelter, and infrastructure needed to promote economic growth and a better quality of life in urban areas.

8

Urbanization and the Environmental Risk Transition

KIRK R. SMITH
YOK-SHIU F. LEE

Cities have two general categories of human environmental risks: Those that directly affect health, such as pollution, and those that may be no less damaging, but operate indirectly by impairing ecosystems (e.g., estuaries) that humanity depends on. Although we focus here on environmental health, we recognize that cities often pose important indirect environmental risks as well.

In 1971, Abdel Omran proposed that patterns of disease and death could be more explicitly incorporated into population theory by adopting a framework he called the Epidemiologic Transition. The Risk Transition we discuss here is an extension of Omran's concept to help explain the dynamics of environmental health risks in cities in the developing world (Smith 1990b). The two concepts share some similar ideas. First, economic development is accompanied by great reductions in some kinds of ill health and increases in others. The historically high "traditional" sources of ill health associated with rural poverty, such as infectious and parasitic diseases, trend downward with economic development, although at varying rates. As traditional diseases decline as causes of death, "modern" diseases like cancer and heart disease take over. One definition of the *transition point* might be the level of development at which there is equal probability of dying by modern and traditional causes (Figure 8.1).

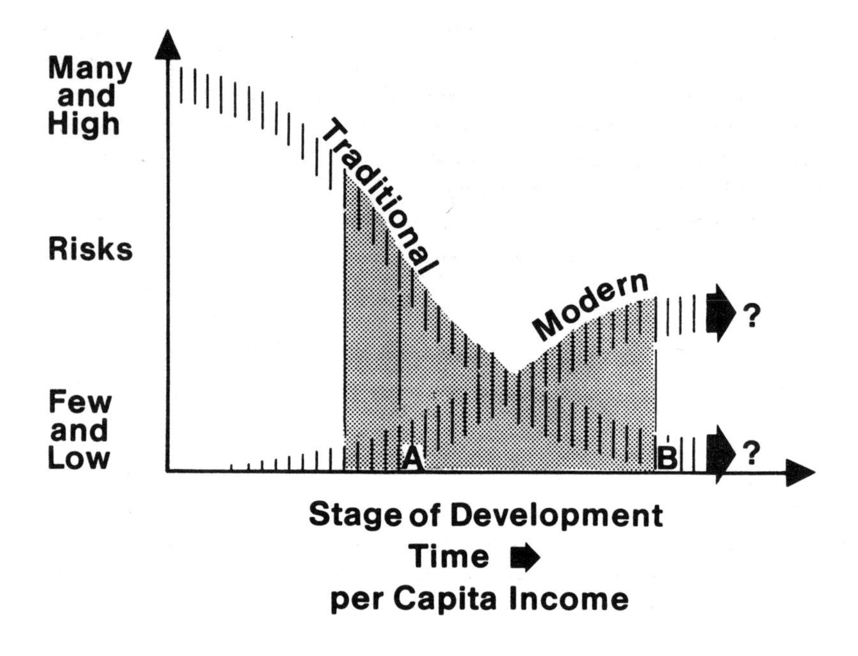

Figure 8.1. The Risk Transition

Because everyone eventually dies, trends in causes of death do not indicate whether overall health improves with development. For this purpose, one can use an indicator of life years lost to various causes of death. Here the transition point might be the level of development at which life years lost to traditional and modern causes of death become equal. Compared to mortality, loss of life expectancy tends to weigh traditional risks more heavily, because of their proportionally higher impact on infants and children. Loss of life expectancy to modern risks does not match that lost to traditional risks until overall life expectancy reaches about 60 years (Preston 1976). This corresponds to an income per capita of $1000 in 1987 dollars—about equal to the average for "lower-middle-income" developing countries (Smith 1990b).

Globally, there has been a dramatic decrease in overall ill health with economic development, at least measured by life years lost, as seen in national, international, longitudinal, and cross-sectional studies of trends in infant mortality, life expectancy, and other indicators of overall health. Indeed, this remarkable drop, driven by the reduction in traditional causes of ill health, is arguably the most important curve in human history. Such net improvement is by no means inevitable, however. Recently released information about Eastern Europe, for example, reveals that its nations were able to control traditional risks rather well but failed to blunt the rise of modern risks to a sufficient extent, causing overall death rates to start rising.

Though the overall pattern is clear, how particular social, economic, and public health changes have induced this reduction is not well understood. Even so, it seems fair to say that cities are where such changes are most concentrated and most rapid. To the extent that urbanization increases, cities would seem to play important roles as bringers of health, although significant numbers of the poor may not experience such benefits (Harpham, Lusty, and Vaughn 1988). Figure 8.2 shows a general relationship between health status and urbanization. Given the dismal environmental health conditions and other stresses in many large cities in developing countries, cities would seem also a primary source for the risk factors leading to modern diseases. We want to address this apparent paradox between bringing and denying health.

The Risk Transition

In contrast to Omran, we shift focus from all forms of ill health to those related to environmental and technological factors—those resulting from people's actions at individual or collective levels as part of economic and other activities. On both sides of the transition, a large proportion of death and disease is environmentally related. For example, traditional infectious diseases and modern cancers are enhanced by such environmental factors as poor water, air, and food quality. If one classifies diet as environmental, as some do, then environmental factors probably dominate on both sides of the transition. For our purposes, the exact environmental proportions—which are difficult to determine and depend on arbitrary definitions and distinctions (see Peto 1980)—are not important. What matters is that the environmental health risks undergo transition as do the patterns of overall ill health.

Also in contrast to the Epidemiologic Transition framework, we shift from a focus on outcome (ill health measured by disease category or cause of death) to a focus on the factors leading to those outcomes (risk). For Omran's purpose of facilitating an understanding of demographic changes, ill-health outcomes are appropriate. To further our understanding

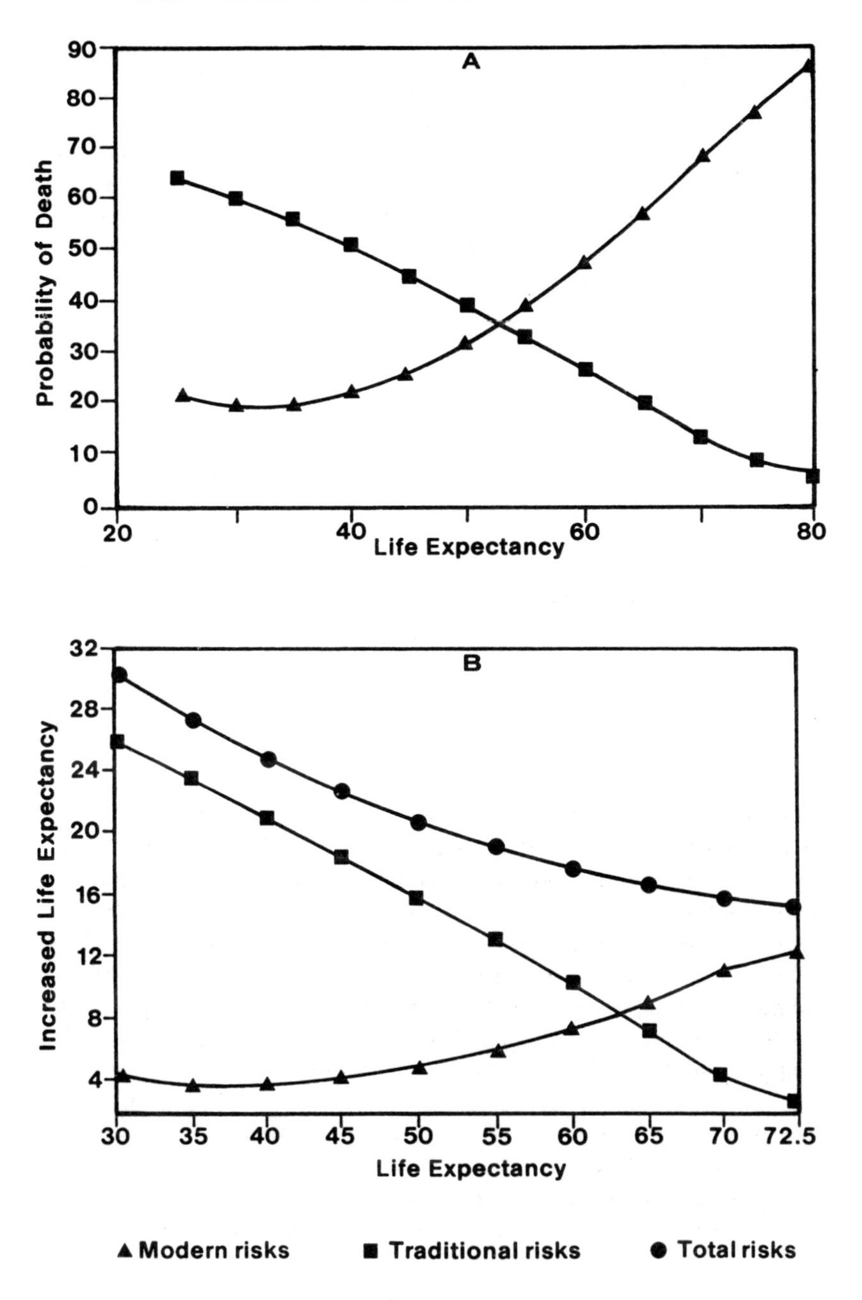

Figure 8.2. Probability of Death and Increased Life Expectancy

of the environmental health implications of specific development decisions, such as different industrialization policies, we need measures closer to the action being contemplated.

To return to Figure 8.1, a modern risks curve would probably be higher than has yet been expressed in modern ill health. The full effects of modern technology, including carcinogenic and climate-modifying pollutants, may not be expressed until many years have passed, but are no less real because of that delay. They have already produced risk, even though there may be considerable uncertainty about exactly how much ill health has been caused. One might even argue that the true modern risk curve rises so high as to cause an upturn in the overall pattern of risk—that at some point additional development increases total risk. This remains to be seen, however.

Before going further, we emphasize that all cities do not have the same risk transition pattern. Rather, we suggest only that the concept of a risk transition provides a useful framework for comparison and analysis of changes in environmental risk experienced in a number of cities in developing countries. There are other aspects of this framework (Smith 1990b), but here we focus on two implications for environmental risk assessment and management: the importance of considering risk overlap and net risk.

The Risk Overlap and Resulting Interactions

At Point A in Figure 8.1, modern risks have begun to rise, but traditional ones are still significant, although rapidly dropping. Rural examples include pesticide runoff added to water pollution caused by poor sanitation and crop residues used for cooking fires in areas of increased pesticide use, creating an entirely new airborne risk in addition to that caused by the smoke itself. If these effects were merely additive, there might be no special need to take note of overlap. Unfortunately, however, risk overlap can lead to other interactions with important implications for risk assessment (Greenberg 1986). For example:

1. *Risk genesis,* where entirely different sorts of risks are created. Urban examples include:
   - Mixing modern (motorized) and traditional (muscle-powered) vehicles leads to new kinds of accident risks and risk management needs.

- Garbage scavenging in dumps, where traditional poor sanitation and modern chemical risks are mixed, leads to hazardous exposures not existing in either traditional or modern societies alone (see below).
- In cities where drinking water supply is still unprotected, uncontrolled release of toxic chemicals can lead to high exposures to hazardous materials.
- Use of modern hazardous materials before implementation of modern regulatory controls produces risks, such as the spread of radioactive material from discarded hospital equipment that occurred in Brazil in 1987.

2. *Risk mimicry,* where morbidity and mortality may be attributed to traditional sources of risk but may actually derive from modern risks.
3. *Risk competition,* where abnormally high or low risks of one disease may actually indicate decrease or increase in the risks of an entirely different disease. Low lung cancer rates in developing countries, for example, may not necessarily mean that the risk factors for lung cancer are lower compared to places with high rates; people may be dying of other diseases with a younger age pattern. This illustrates the distinction between risk and ill health. Use of age-adjusted rates may only compensate for part of this effect.
4. *Risk synergism,* where exposure to one agent causes immunity or sensitivity to other agents in the future.
5. *Risk layering,* where the movement of people concentrates risk in one region and dilutes it in another. This may be an important effect in the many developing countries undergoing significant rural-urban migration where migrants are usually healthier than the average person left in the villages. Thus cities may look healthier than they really are (Williams 1990). Other types of movements, such as by refugees, may have the opposite effect. In addition, it is possible for migration to lower (or raise) the risk levels of origin and destination simultaneously. Also, the origin of risks may be difficult to pinpoint; with chronic diseases or those with long latency periods, the place where the insult occurred may not be where the disease appears.
6. *Risk transfer,* where efforts to control risk from traditional hazards may enhance modern risks (or vice versa), as when pesticides are used to control malaria. Such transfer can also occur from one type of modern (or traditional) risk to another, often involving a shift in the medium, time, place, and people affected, as with air pollution controls that create hazardous solid wastes.

The interactions resulting from risk overlap have many impacts on the demography, economy, and public health of developing countries. Here, however, we focus on the implications for assessing and managing environmental risk. The basic premise is that substantially different risk assessment and management methods are necessary in developing countries because risk patterns are so different from those of the countries where risk assessment methods have been devised—particularly in cities, where much of the interaction between modern and traditional risks occurs.

Net Risk

Consider the different pattern of risks in countries represented by Points A and B in Figure 8.1. Risks at Point B are increasing and in some cases intensifying through modern hazards from further development. The traditional hazards have been mostly controlled and are of less concern, so assessment efforts have been focused on quantifying and managing modern risks.

At Point A, however, important traditional and modern hazards exist side by side, although moving in opposite directions. Introducing technology and the resulting development (interpreted as a movement along the horizontal axis in Figure 8.1) will have a substantially different impact on overall risk in developing countries from that in developed countries. On average, a country at Point A will experience a significant degree of risk lowering (movement down the traditional risk curve) from new technologies as well as risk raising (movement up the modern risk curve).

Although the overall result of development may be general risk lowering, as in Figure 8.1, each particular project will have its own pattern, which may not lower the aggregate risks. Thus, understanding the overall, or net risk from any particular project, regulation, or activity will require evaluation of both classes of risk. Adopting risk assessment methods from developed countries that focus solely on modern risks may be quite misleading.

Development leads to "indirect" routes to risk lowering, versus such "direct" routes as health care. Indirect routes include better jobs, housing, education, communications, and mobility. By comparison, routes to risk raising include the indirect impact of urban stress and the direct effect of chemical pollution. That the risk lowering is indirect need not mean that it is small. A striking example of the effectiveness of indirect routes is provided by a study of all developing countries except China (Preston 1985). Increased life expectancy from a general increase of 1 percent in gross domestic product for these countries was estimated as 2-3 weeks compared to the impact of investing that much income on education—estimated to be 5-20 years.

Of the many studies of the overall relationship of economic development with health status, few have conducted analysis at the project level where individual decisions are made and policies are implemented. Net risk analysis would seem particularly important in Third World cities in understanding the trade-offs between the reduction in risk that projects might bring versus the potentially devastating increase in risk that might result.

The Urban Environment

Here we discuss the principal environmental media—water, air, solid waste, and land—with regard to urbanization and risk transition. We examine present conditions in Third World cities and illustrate how they fit into the risk transition framework. To illustrate the dearth of information that exists for all the environmental media, we focus in the opening discussion on data limitations of urban water supplies.

Urban Water Problems

Water supply and sanitation facilities seem to be the most important aspects of housing design that influence people's health status (WHO 1988b). The World Health Organization (WHO) estimated that 70-80 percent of Third World hospital beds are occupied by patients with waterborne diseases (Wurzel 1989).

Despite their importance and compared with the relative abundance of demographic and economic indicators, information on urban water supply and sanitation services is scanty and plagued by problems of definitions and interpretation. The only major global data on urban water and sanitation services were those WHO collected from public health officials in developing countries and WHO experts responding to questionnaire surveys conducted in 1980, 1983, 1986, and most recently, 1988 (Prost 1989; UNEP 1987; WHO 1988b).

Not surprisingly, the extent of coverage of urban water supply services is positively linked to higher per capita income (Table 8.1) and therefore to urban areas (Table 8.2). These global observations, however, do not fully reveal the exact scale and nature of the problem. Although urban areas have traditionally been better served than rural areas, no reliable statistics are available on the extent of water supply and sanitation coverage in low-income urban settlements. Limited in-

Table 8.1 Access to Water Supply Services: World Population Coverage, 1985-1987

Per Capita Income Categories	*Percentage of Total Population* Urban	Rural
High income (> $5000)—37 countries	100	99
Upper middle income ($1800-5000)—24 countries	90	64
Lower middle income ($500-1800)—46 countries	77	45
China & India ($300)	69	61
Low income (< $500)—42 countries	65	32
Total—151 countries	86	57

SOURCE: Adapted from Prost 1989, 19.

formation indicates that residents in urban slums receive a lower level of such services than their rural counterparts (UNCHS, 1987).

Estimating overall coverage of urban sanitation facilities is made more difficult because definitions vary from country to country and private facilities usually outnumber public ones (Prost 1989). WHO researchers tried to overcome this problem in compiling the global figures by defining access to sanitation services in urban areas as being served by connections to public sewers or household systems—pit privies, pour-flush latrines, septic tanks, and communal toilets (UNEP 1987). Urban sewage disposal systems do not necessarily imply proper waste treatment. In many developing countries, the wastewater collected even by central sewage systems is often discharged into rivers, canals, and coastal water only partially treated or totally untreated.

Similarly, for clarity, WHO defines access to safe drinking water in urban areas as having water piped to a housing unit or to a public standpipe within 200 meters (UNEP 1987). In-house water connection is no guarantee of regular water supply. In Ibadan, Kaduna, and Enugu, Nigeria, for example, 25-34 percent of urban households with in-house potable water connections never had water supplied from their taps (Onibokun 1989). In Bombay, an estimated 63 percent of urban households with in-house connections receive water for 3 or fewer hours each day (Unvala 1989).

Besides masking the problem of irregular supply, the aggregated data on access to safe drinking water fail to reveal much about how much water is actually used by different segments of the urban population. Urbanites with piped water from a communal standpipe 200 meters

Table 8.2 Estimates of Water Supply and Sanitation Coverage in Developing Countries by WHO Regions

WHO Region	*Water Supply* 1980	1988	1990	*Sanitation* 1980	1988	1990
Africa						
Urban	69	77	77	57	79	80
Rural	22	2	27	20	17	16
Americas						
Urban	83	87	87	74	81	82
Rural	41	47	49	11	19	21
Eastern Mediterranean						
Urban	84	89	90	53	76	79
Rural	31	28	27	8	10	10
Southeast Asia						
Urban	67	66	65	29	34	35
Rural	31	56	62	7	12	13
Western Pacific						
Urban	75	74	74	92	94	94
Rural	41	50	52	64	67	67
Global total						
Urban	76	78	78	56	66	66
Rural	31	46	49	14	17	18

SOURCE: WHO 1988b.

from their home (thus having access to safe drinking water) cannot use as much as those with steady, direct in-house supply. Estimates from Habitat suggest that water consumption for urban residents who depend on communal taps within 200 meters of their homes is 20-40 liters per capita per day (Lpcd). Urban residents with a single yard tap may consume 40-60 Lpcd, whereas households with multiple in-house fixtures—commonly in high-income areas—can reach 200 Lpcd (UNCHS 1987).

Thus, the global and regional data are grossly inadequate to a full understanding of the exact magnitude and nature of the distribution or deficiencies of the current urban water supply and sanitation systems.

The combined effects of population growth, industrialization, and urbanization on water supplies lead to four environmental risks: They limit the quantity of clean water for households and the reliability of these supplies; they increase lowland subsidence and flooding and expand ground and surface water pollution. These impacts are exacerbated by the mix of modern and traditional activities (risk overlap) in Third World cities and cannot be assessed accurately without considering interacting effects (net risk).

Water Quantity. Urbanization and associated industrialization can result in overpumping of groundwater, leading to lower water tables and land subsidence and, in coastal areas, saltwater intrusion into the aquifers. This process decreases access to clean water by lowering supply and increasing contamination. Thus, urban residences served by traditional supply systems, such as local wells or groundwater pumping, are at health risk from the rapid introduction of industrial water demand that often accompanies urbanization.

Groundwater is a major source of water supply for some Third World cities. The Bangkok metropolitan region, for example, relies on deep groundwater for about half its total water consumption (Sharma 1986). Two-thirds of Jakarta's inhabitants are supplied exclusively from groundwater. As a direct consequence of groundwater overdraft and seawater intrusion, many wells have been abandoned. The quality of the city's aquifers has been adversely affected up to 8 kilometers from the coast in northern Jakarta. A major problem associated with groundwater use in developing countries is that almost all wells are privately owned and operated by individuals or firms, so groundwater development is unplanned, uncontrolled, and unmanaged. The number of wells and the amount and depth of extraction in Jakarta remain largely unknown (Sharma 1986).

Water Reliability. To counteract traditional risks effectively, household water must not only be available in adequate quantity, but the distribution over time and among different neighborhoods also must be sufficient to meet most people's needs most of the time.

Continuous water supplies at adequate pressures are virtually nonexistent in most urban areas in developing countries. Intermittent supplies of piped water seem to be the norm in most Asian and African cities (Onibokun 1989; UNEP 1982) because of scarcity of needed equipment, material, and skilled personnel. In addition, power outages represent a major handicap for urban water supply systems. The irregular electricity supply causes pumps to shut down and reduces water pressure; it further compounds the problems by damaging the water pumps and water treatment plants. This results in extra costs, because furnishing water intermittently leads to higher peak demands, which larger-diameter and more expensive pipes are necessary to meet (Lindh 1983).

Supply irregularity affects health directly by limiting access to household water, and suction during shutdown greatly enlarges the risk for pollution intrusion into the water distribution system.

Lack of foreign exchange has serious impacts on urban services, including water, in developing countries by reducing access to imported parts and supplies. For instance, in Dakar and Ibadan, water flow may be reduced or intermittent, not because of lack of raw water, but for lack of foreign exchange to buy the chemicals to treat it (White 1989).

Complicating the developing countries' metropolitan water supply problem is their aged, traditional piping systems. Many new water systems are ad hoc additions and alterations to existing facilities with insufficient regard to long-term planning. In early 1984, a new, large World Bank water supply project came on line in Nairobi; water pressure throughout the existing system soared and pipes burst everywhere (Stren 1989). Thus, the combination of old and new worked less well than either on its own.

Old pipes are prevalent in most urban water supply systems in developing countries, and adequately trained personnel and equipment to monitor, detect, and repair broken and leaking connections are often insufficient. Leakage or unaccounted-for water may make up as much as 40-60 percent of the total water supply in Third World cities.

Detecting and repairing leaks in water distribution systems is one of the most vital measures for minimizing costs and maximizing services, but the need is still largely ignored in developing countries. Percentage water losses actually tend to increase as water supply and distribution systems expand. A major recurrent problem is that operation and maintenance of existing systems inevitably receive lower priorities for personnel, equipment, and financial resources than system expansion (UNCHS 1984a).

Land Subsidence and Flooding. Unrestricted exploitation of aquifers can lead to land subsidence. In 1982, groundwater extraction from deep aquifers in Bangkok equaled 1,400,000 cubic meters per day, far exceeding the estimated safe yield of 600,000 cubic meters through natural recharge. As a result, piezometric levels in the wells have been declining rapidly and the ground level has subsided by more than 0.5 meters. Some parts of the city are settling at a maximum of 10 centimeters per year (Sharma 1986).

In addition, as impermeable or near-impermeable surfaces increase through construction activities, infiltration into groundwater is markedly reduced. Storm runoff increases and accelerates, and peak flows swell. Surface runoff from impervious areas may be hundreds of times greater than runoff from some natural areas, and complete urbanization of a watershed may increase the mean flood discharge by five to ten times (Gladwell 1989). Coupled with the effects of soil erosion and sedimentation in rivers and canals resulting from urban construction, flooding in low-lying areas is more frequent.

The Bangkok region, only about 0.1-1.5 meters above mean sea level, suffers from the latter three phenomena (Sharma 1986). Land subsidence, exacerbated by poor drainage and high tides, results in a higher frequency of flooding during heavy and moderate rainfalls, which raises risks of a range of waterborne and water-related diseases, particularly in the lowest-lying and poorest areas. In parts of central Jakarta, land surface has subsided up to 0.8 meters and parts of Indonesia's capital city may be subsiding at a rate of 1-3 centimeters a year (Sharma 1986). Mexico City has sunk 10.7 meters during the past 70 years into the filled-in lake bed on which it was built (Lindh 1983).

In addition to health risks, land subsidence causes structural damage to buildings, roads, railway lines, and underground pipelines. Because land subsidence may reverse gradients, municipal drainage and sewage systems can quickly become obsolete, exacerbating other water-related health risks.

Urban Sanitation. A number of obvious sources pollute water supplies. Septic tanks and other sewage systems, if not properly constructed, located, and maintained, can easily pollute ground and surface waters. Insufficiently treated or untreated industrial and municipal wastes discharged into water bodies pollute water supplies and pose risks to human health. In residential areas served by traditional water supply systems, the risks posed by such pollution are magnified because of direct exposure routes—mainly the effect of poor human waste control, although chemical and other industrial wastes are growing in importance.

Water supply facilities have advanced faster than wastewater management. Although three-fourths of urban dwellers in developing countries had water supply facilities in 1988, only two-thirds had sanitation services (Table 8.2). Ironically, the higher level of development of urban water supply facilities may adversely affect the urban environment, because increased water supply may lead to unhealthy situations if wastewater disposal systems are inadequate.

Developing countries' major sources of water pollution are untreated or partially treated domestic sewage, industrial waste effluent, and domestic and industrial garbage. Bangkok, Jakarta, and some smaller Indonesian cities currently have virtually no public sewage disposal system (Lindh and Niemczynowicz 1989). In Bangkok, human wastes are deposited in septic tanks, cesspools, and then storm water drains. Such a sewage disposal method has health risks in a city where an inefficient drainage system combines with periodic flooding to pollute water sources. The minimal role of central sewer systems in urban Africa is also evident. In Dar es Salaam, fewer than 10-15 percent of households were served by the central sewage system in 1983 (Stren 1989). In Khartoum, the figure for 1986 is 5 percent (El Sammani et al. 1989).

China, with over 200 million city dwellers in 1980, had only 35 small municipal wastewater treatment plants, so 80-90 percent of the estimated 37 billion cubic meters of sewage discharged into water bodies throughout the country is emptied without any treatment (Liang 1989; Smil 1984). In Shanghai, China's leading metropolitan region, only 4 percent of the estimated 5 million tons of wastewater discharged in 1980 was treated. In 1979, 96 percent of the surface water test points in the Shanghai municipality were contaminated with heavy metals (Smil 1984). The heavily polluted water may contribute to the rapidly rising cancer morbidity and mortality in China's leading industrial region, but scientific evidence to support such a claim is not yet reported.

In the Association of South East Asian Nations (ASEAN) region, except for Thailand, the small size of river systems greatly limits their capacity to accommodate and dilute urban and industrial wastewater. For example, in Kuala Lumpur, because of heavy water demand from urban and industrial activities, water stored in two dams upstream in the Kelang River Basin is rarely released. As a result, extremely low flows downstream of the dams are unable to dilute and flush the liquid and solid wastes generated by the urban centers. Kelang River has been converted in effect into an open sewer (Low 1989).

Urban Air Pollution

Although insufficient environmental or medical monitoring data are available, it appears that urban air pollution in developing countries may be a major health hazard, particularly for the old, young, and already ill. Results from the Global Environmental Monitoring System (GEMS—part of the United Nations Environment Programme and WHO), though covering only a few cities, show that substantial portions of the cities in developing nations have outdoor pollution levels that substantially exceed health standards (GEMS 1988). Although there are many risk factors other than air pollution, respiratory diseases are the major cause of death and illness in developing countries (WHO 1986).

There has been a recent shift in the focus on air pollution. Some pollutants produce damage on very large scales, such as the regional impacts of acid rain and global impacts of greenhouse gases. But the health impacts of other air pollutants can only be understood through careful consideration of what occurs in such microenvironments as vehicles, homes, and workplaces. In either case, the traditional focus of air pollution monitoring and regulation at the intermediate level of urban outdoor air quality is no longer considered adequate by itself (Smith 1988a).

Although air pollution is associated in many people's minds with industrialization, urbanization, and fossil fuels, and thus is a modern risk in the framework of Figure 8.1, the actual risk pattern is more complicated. Risks from some health-damaging pollutants, such as combustion-generated particulates, actually seem to decrease with development—making these traditional risks. This situation can occur even when ambient (outdoor) urban concentrations are rising because of increased use of fossil fuels. Exposures, which more closely indicate health impacts, can undergo rapid decreases during the same period.

This apparent contradiction arises because exposure is a function not only of environmental concentrations but also of the number of people affected and the time involved. Thus, if concentration is measured in parts per million (ppm) of chemical in the air, an appropriate exposure measure might be ppm-person-hours—the number of people breathing air of a certain concentration for so many hours. This emphasis on exposure is simply stating that if one is worried about ill health caused by air contaminants, it is necessary to measure air where the people are. This seems obvious, but environmental monitoring systems in most countries have been focused on ambient concentrations measured at convenient, secure, outdoor locations (e.g., post office roofs) where people hardly ever are.

Only a small percentage of the world's person-hours are spent where most air pollution monitoring, regulation, and research is focused—outdoors in the cities of developed countries. The same pattern is evolving in developing countries. Everywhere in the world, a much larger fraction of time is spent indoors. Although indoor pollutant levels will be influenced by outdoor sources (and vice versa), recent research shows that outdoor concentrations are generally poor predictors of total exposure, being too high or too low, depending on the situation and pollutant (Smith 1988b).

Although indoor air pollution is also seen as a developed country problem, it may actually be much more important in developing countries.

In particular, more than half the world's households still cook with solid fuels: in probable order of importance, wood, crop residues, charcoal, coal, and animal dung. These fuels produce rather large amounts of pollutants per unit of fuel. Although unprocessed solid fuels are more prevalent in rural areas, large numbers of the poorest urban dwellers also rely on them. Because of the poor ventilation and the location and timing of these daily emissions in each household, total human exposures are much higher than might be indicated by looking solely at emissions from household fuel use (Smith 1988b). Risks from household fuel use follow traditional patterns. Moving up the energy ladder, households shift to higher quality fuels with development. They are more expensive but can be used more efficiently, conveniently, and cleanly (Smith 1990a).

GEMS (1988) has tracked outdoor pollution in several dozen cities across a range of incomes. Emission trends in the two best-studied pollutants, sulfur dioxide and particulates, are mixed; some developing nations show slow downward trends (China), whereas others show significant (India) or spectacular (Thailand) increases. Even leaving aside the uncertainty of such estimates, the shift in temporal and spatial patterns of these emissions makes it difficult to draw conclusions about potential health impacts. For this purpose, exposure measures are preferable.

As with indoor air pollution, distribution of outdoor concentrations by level of development may not fit preconceived notions about the relationship of industrialization and environmental contamination. GEMS data show that poor ambient urban air quality, as measured by ambient particulate concentrations, seems to be a traditional risk today (tending to decrease with development).

Solid Waste

Solid wastes dumped into rivers and canals overburden the water's capacity to dilute and flush the waste materials. Left in overfilled and underdesigned waste disposal sites, decomposed solid wastes can easily pollute groundwater through seepage, particularly in the humid tropics. Much of the municipal garbage in developing countries is disposed of by open burning, dumping into rivers or canals, and dumping into abandoned mines and swamp areas. In the Bangkok metropolitan area, an estimated 50 to 100 tons of garbage are dumped each day into the city's rivers and canals (Low 1989).

According to Whitney (1991), the way cities and their surrounding areas are connected by resource and waste flows tends to go through a transition during development. In traditional societies, the city uses resources from the immediate rural areas and urban wastes (*unused outputs* in Whitney's terminology) are essentially entirely used in nearby rural areas. Evidence of this can be seen in China, where "fertility migration" to agricultural areas directly adjacent to the city resulted from increased use of night soil and other urban wastes.

As development occurs, however, this relationship shifts for several reasons to a situation in which rural areas use comparatively little urban waste: A shift upward in the relative value of labor versus materials makes waste recycling less attractive; a shift downward in the perimeter/area ratio as cities grow larger results in the production of more waste than nearby rural areas can use (transport costs always having limited the radius of potential use); and changes in the composition of urban waste make it less easily used by rural dwellers. To Whitney (1991), when a city still uses outputs of the rural area but no longer produces usable wastes in return, it is a sign of the collapse of traditional society. The result has been a ring of open dumps around cities. For example, such dumps cover 6.5 square kilometers in more than 5000 separate sites around Beijing (Whitney 1991). In Third World cities with more controlled dumping, similar land areas are involved, but in only one or a few locations in or near the cities (e.g., the ASEAN capital cities; UNCRD 1989).

The people who pick, process, and pack municipal waste have, through the centuries, presumably been exposed to a range of hazards, particularly infectious agents associated with the human excreta often found in such wastes, as well as with the generally unsanitary conditions (Blincow 1986; Taira 1969). Shifts in waste composition through modernization and industrialization have not only led to changes in the value and potential uses of municipal waste because of a higher fraction of ash, plastic, and inorganics in general, but also to entirely new types of risks for these people.

Picker (sometimes called recycling or scavenger) groups in industrializing cities are now exposed to new contaminants from the city's modernizing sector as well as the traditional risks of their occupation (Furedy 1989). The hazardous chemical and hospital wastes found in municipal refuse may lead to exposures entirely different in quality and quantity from those found in either completely traditional or modern situations (Kungskulniti et al. 1991). The urban poor in general live at the interface between the modern and traditional sectors (WHO 1988a), but garbage pickers would seem to be an extreme example of risk genesis through the risk overlap described above.

Municipal waste treatment in mixed economies often leads to risk transfers as well. Air pollution from simple incineration and spontaneous combustion as well as water pollution from leaching are common accompaniments of urban landfills in developing countries. Although improperly handled solid waste can have serious adverse health consequences, solid waste management—much like wastewater treatment and disposal systems—frequently receives a lower priority than other municipal services. A major reason for this most serious shortcoming of public works is that waste disposal has historically been relegated to the lowest levels of responsibility (Mehta 1982). A related problem is the unavailability of suitable landfill sites because of high land costs and the unplanned nature of rapid urban growth in developing countries.

Urban Slums

The potential for interactions resulting from risk overlap is greatly affected by urban land use, whether planned or unplanned. Risks operating through all of the environmental media can be involved. We give examples below.

The urbanization process has often been haphazard and chaotic in developing countries. The ill effects on health resulting from urban water and air pollution have been felt disproportionately by low-income urban residents, who have no alternative to overcrowded slums and shanty towns. These low-quality, occupant-built areas are usually only marginally served by the municipal utilities others take for granted.

For example, poor urban residents often lack a potable water supply and adequate resources for removal and disposal of household and human wastes. Moreover, the nature of their dwellings often makes them more vulnerable than others to such hazards as dust, chemical pollution, and noise.

Although unavailability of uncontaminated water is a problem in most developing countries, unplanned and rapid urbanization appears to be increasingly associated with the deterioration of existing water systems. In densely populated areas where facilities for household and industrial waste disposal are inadequate or nonexistent, the bacterial contamination of water by industries and human wastes is becoming a major problem.

Overcrowding, poor ventilation, lack of piped water, a high density of insects and rodents, lack of waste disposal facilities, poor drainage, nearby industrial facilities, and heavily traveled roads—all typical of the urban slum and squatter environment—are predisposing factors for a range of traditional and modern diseases. There seem to be strong associations among housing, land use, and health risk.

Conclusion

If cities are so unhealthful, how can they bring health? The risk transition helps resolve this apparent paradox by pointing out the need for net risk assessment—the examination of risk lowering as well as risk raising. The risk transition also shows that the net risk result can change over time or with location. Because of the time delays inherent in many modern risks, it is possible to be fooled by the difference between committed and expressed risk—ill health. Committed but yet-to-be-expressed risks may start to mass faster than the traditional forms of ill health decline—and then the city stops improving health even if overall ill health still seems to be declining. Identifying this point should be a major focus for cities in developing countries, where this event is likely to occur first.

Although not discussed here, Third World cities may be ecological saviors as well as ecological disasters because they offer one of the few proven routes for reducing pressure on sensitive and disappearing natural ecosystems and resources, such as forests. They may create many problems faster than they solve them, but they also provide some solutions to ancient dilemmas.

Costs and benefits of development and urbanization, in environmental health as elsewhere, are unequally distributed—not every group will experience a net negative risk, even if the overall risk is negative. The urban poor are more luckless than others, to paraphrase Eckholm (1977), as they face simultaneously the health risks generated by their own poverty and by other people's wealth. With limited or no access to housing, water supplies, and sanitation, they are constantly exposed to the ill effects of infectious diseases. At the same time, they must also contend with the toxic effluents of affluence associated with industrialization and urbanization.

A continually larger fraction of humanity will be found in cities, making it more important to find effective ways to manage the urban environmental risk transition. At a minimum, this will require choosing indicators and decision-making frameworks appropriate for making trade-offs in a rapidly changing pattern of falling and rising risks with critical distributional implications across generations as well as income groups.

9

Transnational Capital Flows, Foreign Investments, and Urban Growth in Developing Countries

VICTOR FUNG-SHUEN SIT

Since the early 1980s, urbanization studies in less developed countries (LDCs) have taken on new dimensions and approaches quite distinct from those of the 1960s and 1970s. A new international division of labor unfolded (Cohen 1981; Friedmann and Wolf 1982) and successes and failures of the two major models of LDC industrialization strategies—import substitution industrialization and export-oriented industrialization—provided subjects for more globally oriented studies (Dunning 1988; Sanderson et al. 1987; Taylor and Thrift 1982). Even in studies that are only national in scope, external factors have been introduced to illustrate how most countries are closely linked through trade and investments and that urbanization as a social process reflects such linkages (Abumere 1982; Arn 1987; Forbes and Thrift 1987; Gwynne 1985; Rogerson 1982).

Socialist- or Marxist-oriented scholars, who were quite active by the early 1970s, shared the transnational approach (Castells 1972; Harvey 1975; Roberts 1976), although their world system rested on different underpinnings—capitalist accumulation and, particularly for Third World countries, increasing dependency on industrialized capitalist core countries (Armstrong and McGee 1985; Frank 1966; Friedmann 1986).

The implication of the new lines of study for policy is also significant. Some believe LDC governments can adopt policies that let them actively take part in the world economy while ensuring positive feedbacks to their cities (Kasarda and Crenshaw 1991; Teune 1988); they see foreign investment and other international capital flows as having positive impacts on national urbanization processes. Others, like Portes and Johns (1986), are pessimistic, seeing little escape from dependent urbanization for LDCs in the new global order.

Here I describe and assess two decades of transnational capital flows and their implications for urbanization. I also discuss problems of concepts, data sources, methodology, and the general inadequacy of studies relating urbanization to international capital flows, concluding with implications of work to date for policy.

Forms, Nature, and Volume of Transnational Capital Flows

It is useful to begin by reviewing the types and size of flows of financial resources from industrial to developing countries (Table 9.1). In the early 1960s, official development assistance (ODA) provided the major flow, mainly with resources from developed country governments and multilateral agencies to promote LDC economic development and welfare. ODA increased less rapidly than the total flow in the 1960s and declined as a percentage of GNP for the LDCs, from 0.52 in 1962 to 0.35 in 1971 (OECD 1973, 6). But since the mid-1970s, it has grown more rapidly. There is also a steady, increasing trend in the ratio of loans to grants. From 1961 to 1971, the proportion of loans increased from 21 to 49 percent of total ODA flows.

As most of ODA is concentrated in physical and social infrastructural improvements of the host countries (with restricted amounts going into projects directly related to promoting economic activities), and good data on its urban-rural split are lacking, little is known of ODA impacts on urbanization in the host country. It is generally felt, however, that through ODA, developing countries have been able to influence policies of LDCs, including those related to urbanization (Lubell 1984).

Other official flows consist of those from the public sector on terms approximating market conditions—for example, official export credits and central banks' taking up bonds. Such flows have steadily increased over the years and proportionately account for a larger share of the total flow. Between 1969 and 1982, these flows averaged $6.4 billion annually—27 percent of the total flow. But these flows are largely aspatial in nature.

Table 9.1 Net Flow of Financial Resources From Industrial to Developing Countries, 1960-1982 ($US billions)

	Averages			
	1960-1966	*1967-1973*	*1974-1979*	*1980-1982*
Official Development Assistance	5.5	7.4	16.2	26.9
Other Official Flows	0.5	1.2	3.4	6.4
Private Flows				
Direct Investment	1.8	4.3	8.7	12.0
Export Credits	0.7	2.1	6.7	9.8
Others*	0.7	2.1	14.6	25.6
Grants by Private Voluntary Agencies	—	0.8	1.5	2.2
Total	9.2	17.9	51.3	83.0

SOURCE: International Monetary Fund 1985.
*Excluding short-term flows.

Private flows at market terms include four major elements: direct investment, portfolio investment, private export credits, and commercial bank loans. Direct investment, as contrasted with portfolio investment, has generally been considered the distinguishing characteristic of multinational corporation investments. Direct investors are "in a position to derive benefits over and above the property income accruing on their capital, whereas portfolio investors have no significant control over the operations of the enterprises in which they invest, and are mainly interested in the value of and returns on the capital invested" (UNESCAP/CTC 1986, 12). Net foreign direct investment (FDI) includes equity capital flows, parent company loans, and the foreign direct investor's share of reinvested earnings.

Private export credits are often promoted by providing rediscounting through central banks and government guarantees, covering up to 90 percent of the invoice value, and debt-creating bank loans. The increased role of banks and financial intermediation in private flows reflects changes in the international financial system spurred by higher oil prices and the buildup of substantial short-term deposits by the principal oil-exporting countries.

Private flows accounted for 34.8 percent of the total flow in the early 1960s; by the beginning of the 1980s, their share had increased to 57.1 percent (Table 9.1). Short- and medium-term commercial credits are now the predominant form of private flow. Grants by private voluntary agencies (foundations, missions, and nonprofit organizations) contribute a minor fraction of the total flow.

Although the amount of foreign direct investment directly connected to multinational corporation (MNC) activities looks small in the total flow, its significance for the host countries' development and urbanization far exceeds its share in the total flows. Official-sector flows are largely aspatial, and although some have policy strings attached that may affect urban or rural developments, it is difficult to document them. Third World governments undertook much of the expansion in borrowing from banks to finance balance of payments or fiscal deficits, or for state enterprises to finance investment programs (International Monetary Fund 1985, 1). The latter, though likely having spatial implications, offers insufficient detail for analysis.

Most discussion and studies on spatial and sectoral impacts of transnational capital flows therefore concentrate on foreign direct investment. Yet an Organisation for Economic Cooperation and Development (OECD 1973) study cautioned against using foreign direct investment alone to cover MNC activities in developing countries. In the early 1980s, various new forms of international investment played an increasingly important role (Oman 1984, 12), such as (a) joint international business ventures in which foreign-held equity does not exceed 50 percent and (b) international contractual arrangements involving at least an element of investment from the foreign firm's viewpoint but possibly no equity participation by that firm. Such arrangements frequently occur with licensing agreements and management, service, and product-sharing contracts and occasionally with subcontracting and turnkey operations.

The new forms of investment constitute a gray area between the classic international activities of firms, namely wholly or majority-owned foreign direct investment and exports. Globally, the new forms of investment have gained importance in manufacturing industries over the last decade, promoted by several factors. Endogenously, host government regulations favor reduced or nonequity forms of foreign direct investment. Structural changes in these countries also create new opportunities and reduce costs in doing business via the new forms of investment. Exogenously, various factors are included: (a) the relative stagnation of growth in the industrialized countries, (b) a diminishing gap of technology separating the industrialized countries from developing ones, (c) instability of international currency exchange rates, (d) increase and instability of interest rates, and (e) a globalization of oligopolistic interfirm competition in numerous key industries.

One result of the new forms of investment is capitalization of important assets of the parent MNC as well as the traditional sectors of economic activity of the host country, hitherto un- or underdeveloped. Through new forms of investment, MNC influence has spread wider and deeper than the traditional measures (Table 9.1).

Foreign Direct Investment Characteristics

The major source of foreign direct investment is the Group of Six (G-6)—the United States, United Kingdom, Japan, Germany, Canada, and the Netherlands. In 1985, shares in offshore manufacturing amounted to 17.3 percent for the United States, 3.9 percent for Japan, and 19.3 percent for West Germany. These industrialized countries in 1987 contributed 81.9 percent of the total world stock of foreign direct investment (Table 9.2). Other characteristics of such investment from these countries may be discerned. Since 1980, outflows of foreign direct investment from Japan and Germany have risen spectacularly, much above the world average as well as the rate of increase of the other four G-6 countries. Their outflows will likely continue as the stock of the respective foreign direct investment of these countries as a ratio of their GNP is still much smaller than that of the other G-6 countries. Pressure on Japan to increase the outflow has been mounting since the early 1980s because of unbalanced trade between Japan and the United States and the yen's large appreciation.

In the 1960s, foreign direct investment from developed countries into LDCs was concentrated disproportionately in projects related to natural resources exploitation, particularly mining and petroleum. That trend has since been altered by nationalization programs in most countries, such as the OPEC countries for petroleum. In the 1980s, there was a clear tendency of foreign direct investment flow toward manufacturing. The regional and sectoral breakdown of this investment flow from developed to developing countries substantiates this pattern (UNESCAP/CTC 1986). The high concentration of such investment in manufacturing in Latin America and the Caribbean in the years 1965-1972 was related to the import substitution industrialization of the major countries in that area and period. In general, in other developing regions, foreign direct investment concentration in manufacturing is a more recent trend.

Table 9.2 World Stock Value of Overseas Direct Investment and Growth Trend

	Stock ($US billions)		*Growth Rate ($US-dominated increasing ratio)*				*Share in Total Flow*	
Major Source Country	*1985*	*1987*	*1984-1985*	*1986-1987*	*Average 1970-1982**	*1984*	*1987*	*1985*
United States	232.7 (35.9)	308.8 (32.1)	9.2	19.0	9.8	13.6	24.6	5.8
United Kingdom	116.9 (18.1)	177.8 (18.5)	16.2	30.8	8.6	11.9	21.0	25.7
Japan	44.0 (6.8)	77.0 (8.0)	16.1	32.5	20.6	17.8	9.5	3.3
West Germany	52.4 (8.1)	100.0 (10.4)	43.2	33.5	17.1	9.2	12.6	8.4
Canada	33.5 (5.2)	46.1 (4.8)	6.0	9.5	8.5	12.6	3.3	10.1
Netherlands	55.5 (8.6)	78.8 (8.2)	37.0	37.5	7.6	7.7	10.8	44.7
6-country total	535.0 (83.0)	788.5 (81.9)	16.3	26.1	NA	72.8	81.9	–
World total	644.6	962.8	16.3	26.1	10.5	100.0	100.0	–

SOURCE: Compiled from JETRO 1987, 1989.
NOTE: Percentage of stock in parentheses; world total = 100%.
*International Monetary Fund 1985 world figures include industrial nations only.

Distribution of foreign direct investment stock within manufacturing (Table 9.3) in Hong Kong, Malaysia, and Singapore underlines the leading role of labor-intensive firms, particularly electrical and electronics. Foreign direct investment activity partly accounts for the increased Third World share in total world exports in the 1980s in such products as communications and office machines, household equipment, textiles, and clothing (JETRO 1987).

Several recent patterns have been observed in foreign direct investment in LDCs that began to quicken after 1986. After 1986, developed countries' foreign direct investment had aimed first at newly industrializing countries (NICs) in Asia, then at Association of South East Asian Nations countries and China. The next waves of foreign direct investment extended to Sri Lanka, Turkey, Chile, and Mexico. Many LDCs have also been trying hard to facilitate incoming foreign direct investment by easing foreign capital control laws and reducing their debt burden through debt-equity swap schemes or other diversified financial measures based on international coordination (JETRO 1986). Thus, the general climate in LDCs is becoming more conducive to transnational capital flows from developed countries, and this flow has gained in volume since 1986.

Role of Transnational Capital Flows

Views on the role of transnational capital flows in LDC economic development are contradictory. Some view its impact as negative, believing that foreign capital reduces host country economic growth, leading to "decapitalization" (Bornshier 1980; Chase-Dunn 1975; Vernon 1971). At the same time, industrial MNCs in LDCs are seen to be committed to a development strategy in which capital is sent mainly into technologically advanced, highly concentrated sectors and industrial branches. This strategy has absorbed much of the capital available in LDCs for investment, so that in branches and regions not penetrated by MNCs, the scarcity of funds for capital formation might be intensified. MNCs compete with LDC national governments in the international money market with greater ability to raise finance than many LDC governments. Because of their low credit rating and the high interest rates demanded as a result, some LDCs have not obtained desperately needed loans (Taylor & Thrift 1982).

Table 9.3 Distribution of Stock of Incoming Foreign Direct Investment in Manufacturing by Industry (in percentages)

Industry	*Hong Kong 1982*	*Malaysia 1979*	*Singapore 1978*
Electric & Electronics	42.8	8.5	34.2
Textiles & Garments	11.6	16.0	8.4
Construction Materials	10.4	NA	NA
Watches & Toys	6.7	NA	NA
Chemicals & Rubber	6.4	23.0	9.3
Food & Beverages	5.1	25.0	7.0
Transport Equipment	2.0	6.9	14.3
Metal	3.3	5.8	6.7
Wood & Paper	NA	5.1	8.8
Total FDI ($US millions)	1260	1203	1208

SOURCE: UNESCAP/CTC 1986, Annex Tables 4, 6, 8.

The Latin American experience illustrates transnational capital flows' negative impact in blocking domestic growth. After a decade of growth, the Latin American economy suffered a reversal. Average per capita income growth decreased in 1980-1985. Net capital outflow totaled a colossal $109 billion. Whatever real aggregate gains may have occurred in the 1970s have as a result of debt buildup since been erased by recession and debt adjustment (Dietz 1986). Latin America's gloomy 1980s economic scene was felt to relate to the dominant role of foreign capital and its integration into national economies because national capitalists of the respective LDCs have been displaced and their growth blocked. Exports from subsidiaries of U.S.-based MNCs have often been banned or severely restricted so as not to compete with the parent or other subsidiaries. These restrictions have frustrated the potential for a manufacturing export stage in Latin America. Such a situation has nothing to do with Latin America's level of protection or overvalued exchange rates, but rather is due to internal policies of MNCs conceived to protect their market shares and profitability and to avoid competition that directly threatens their worldwide source and distribution networks.

Studies in Africa have also identified a high concentration of foreign direct investment and negative impacts on the host economy. Abumere (1982) revealed that in Nigeria in 1963, MNC investment represented 87.3 percent of all private investment and 79.9 percent of all investment, with

60 percent of the investment concentrated in oil and natural gas. Import substitution industrialization is of the most interest to MNCs. Few MNCs in Africa produce capital goods, but almost all produce consumer wares—largely financed by indigenous capital or the host government. Mabogunje (1978) studied MNC activities in Africa, concluding that MNCs import too little advanced technology and that the technology they do transfer into the host country has little chance of being internalized by the indigenous technology. Worse still, MNC activities harm local industries by eliminating their markets, blocking backward linkages through the use of imported raw materials and intermediate goods and thus frustrating the very principle of import substitution industrialization. Also, MNCs concentrate in high-profit industries unrelated to development needs of the host economy, such as beer and tobacco manufacturing, and use transfer pricing to increase their profits.

Dependent industrialization under the sway of foreign capital and technology is likely to lead to premature and high levels of industrial concentration and early symptoms of stagnation in total output (Bornshier 1980). In many LDCs (e.g., Argentina, Brazil, Colombia, Mexico, India, and Indonesia), more than half of the largest industrial enterprises are under foreign control. This high concentration is likely to lead to monopoly and extra profits. In Bornchier's statistical analysis of 75 LDCs with per capita income of $3000 or less in 1967, there is support for the decapitalization thesis. MNC penetration reduces the industrial and regional balance of the host economy and in the long run reduces the funds available for investment by domestic firms. Based on data from 73 LDCs in 1960-1970 and 1970-1980, R. D. Singh (1988) achieved a similar result: MNC economic penetration per se has been of little or no consequence to economic growth in LDCs and the presence of MNCs does not seem to have significant effects on industrial output growth.

Of course, insolvency in a number of Latin American and other LDCs is also the product of the uses of borrowed funds with substantially limited repayment possibilities. In addition, these limited funds are often spent on military expenses or the import of consumables to support OECD country exports.

Another perspective on the role of transnational capital flows in LDCs is growth theory. Papanek (1973) sees private foreign capital inflow having the same role as domestic capital on economic growth, whereas foreign aid exercises a much larger effect than private flows on LDC growth. Kobrin (1977) believes that foreign investment would intensify the relations between social modernization and industrialization.

Although most studies with this perspective agree on transnational capital flows' positive role, the authors also seem to concur that the policies of the host government are critical in affecting the final outcome. For example, Lo, Salih, and Douglass (1978) note the importance of foreign capital in two of their four models of development in Asia, concluding that from the LDC point of view, availability of capital is critical for development paths. It could be seen as an opportunity, but the pivotal variable seems to be the individual nation's capacity to manage domestic policies to maximize benefits from its position in the new international system.

The fear of monopoly and extra profits of MNCs has been the focus of studies under this perspective as well. Kindleberger (1974) offered a notion of *bilateral monopoly* to explain that although MNCs may monopolize certain factors of production, they are also subject to regulation by the state in LDCs that monopolizes the terms of access of MNCs. Moran (1985) also allays the fear of MNC control in asserting that international competition among MNCs is increasing. As MNCs create and expand sales of a product, technological barriers to entry would fall and new foreign and then domestic firms would enter the market. In short, most MNC markets pass through a relatively rapid cycle of monopoly and oligopoly that ends in workable competition that limits the monopoly earnings and the "control" components of MNCs in a developing country.

The pro-transnational capital flows group includes those following the business school and traditional economic approaches. They see these flows as additional investment income and emphasize the role of MNCs as pacesetters in the transfer of capital, technology, modern management, and market information, but largely disregard the political and social impacts. LDC indebtedness, especially in Latin America, is regarded as a temporary result of business cycle vicissitudes and ill-advised public policy.

Spatial Impact on Urbanization

Few studies on transnational capital flows refer specifically to their spatial impact, let alone offer detailed spatial analysis. There are general references to the urban bias of foreign direct investment and some macroquantitative statistical correlations between it and the level of urbanization among selected groups of LDCs. There are also discussions of transnational capital flows in theories positing world cities and a world hierarchy of cities.

Spatial implications of foreign direct investment are viewed primarily in terms of (a) spatially uneven regional development and (b) transformation of national and local labor markets through such processes as migration, deskilling, dualism, and reorientation of managerial loyalties away from the state toward the corporation (Taylor and Thrift 1982).

The notion that foreign investment correlates with urban growth is found in modernization theory. Hoselitz (1960) views urban growth as an inevitable result of development as periphery countries adopt traits of modern nations through industrialization and diffusion of a modern orientation. Cities, too, offer overwhelming advantages for profitable absorption of investment capital by allowing the widest access to the domestic market. Cities facilitate capital absorption through the inherent infrastructure of transport; communications; public utilities; a concentrated pool of labor, cultural, and recreational facilities; and easy access to the seat of national political power and decision making (Crenshaw 1991; Kasarda and Crenshaw 1991; Mattos 1982). Thus, major Third World cities are natural locational choices for MNCs that in turn reinforce spatial concentration.

Armstrong and McGee (1985) view the metropolitan centers of the LDCs as places where a state-MNC-local oligopoly forms and colludes to ensure success of the modernization growth model. Primate cities and central subsystems play a crucial role in these social and economic changes. They are the home base of the collaborating ruling class, which imports the ideologies, behavior patterns, and production of the corporate sector. Elites diffuse their tastes to the rest of the society, so that perceived needs of the poor are reshaped and limited family resources are spent on modern goods. Regional cities, market towns, and rural areas, in turn, are persuaded to buy goods produced in the major cities or imported from overseas.

Growing international investment in manufacturing in Asia reinforces the position of large primate cities (Fuchs, Jones, and Pernia 1987). In Indonesia, Forbes and Thrift (1987) found investment concentrated around Jakarta, where all major domestic and foreign corporations set up their headquarters. Manufacturing is more dispersed toward the periphery but still within the city's ambit, increasing its primacy. Decentralization policies (free trade zones, industrial estates, and transportation policy) have had little impact.

Asian countries following import substitution industrialization involving assembly of imported components experience concentration of such industries and thus growth of the large cities where the labor and market of the growing middle- and upper-income classes are. In the 1950s and 1960s, Malaysia and the Philippines were active in such industrialization and reported such an urbanization pattern as well (Armstrong and McGee 1985).

Despite there being no official locational guidelines for MNCs and foreign investment, in Nigeria a clear urban bias of MNC location has been observed (Mabogunje 1978). In 1969, half of Nigeria's MNC industrial investment was located in Lagos, which with other coastal towns, accounted for 70 percent of the investment. Although MNC investments declined significantly away from the coast, indigenous investments declined only moderately, and state government investments increased with distance from the coast. The major factors of MNC locations are knowledge of the location and economic environment, a distribution network, and information-gathering costs, which largely explain their implicit spatial bias. A marked concentration of foreign investment in the national core region and more generally in the metropolitan areas as a whole was also found for South Africa (Rogerson 1982).

Metropolitan concentration of foreign investment is also noted in Latin America. Before the socialist government took power in 1970, big companies and the modern sectors in Chile were dominated by foreign capital that had invested in 63 of the 100 largest companies. Santiago's spatial concentration was strong. In 1967, it contained 57 percent of Chile's manufacturing employment as well as the more technically advanced sectors, and hence companies with mostly foreign capital were quite concentrated in the primate city (Gwynne 1985). In the post-Allende period of 1973-1980, a certain degree of decentralization took place, but Santiago still had 38 percent of Chile's industrial output and 56 percent of its manufacturing.

Some major metropolises in Latin America, like São Paulo, have become important mediating points in integrating the domestic economy and the international market. São Paulo contains the management and production facilities of large MNCs and state and private national corporations, plus a diverse labor force, which establishes it as a Latin American regional headquarters in the world economic system. The accelerated urbanization that stems from such economic growth led to urban growth rates double that of the population growth rate in most Latin American countries and the emergence of high industrial primacy of metropolises such as São Paulo—described by Kowarick and Campanario (1986) as a new pattern of accumulation. It produced sustained GNP growth of 7.1 percent per year for São Paulo in the 1950-1980 period, and high economic primacy—36 percent of the manufacturing work force, 46 percent of total national wages, 40 percent of industrial value added and industrial investment. Other parts of Brazil increasingly became subordinate primary goods producers.

Kentor (1981) made one of the first cross-national examinations of the impact of transnational capital flows on urbanization in LDCs. In his analysis of 37 LDCs, he found investment dependence (measured by per capita debits on investment income, export partner concentration, and export commodity concentration) to have a lagged positive effect on urbanization in these countries. Investment dependence stimulates growth in the tertiary and informal sectors and inhibits growth in the industrial sector. Kentor concludes that it generates overurbanization in peripheral countries as rural-urban migration continues in spite of a lack of opportunities in formal sectors because of adverse rural conditions and the pull of urban possibilities—the so-called Todaro effect. This factor is exacerbated by the growth of a visibly affluent middle class as a consequence of MNC activities.

Growth of World Cities

Another approach relating LDC urbanization and transnational capital flows derives from MNC and dependency theorists, whose thinking converges on the notions of world cities and a world hierarchy of cities.

Hymer (1975) describes a hierarchical model of MNC organization, laying a foundation for a theory of the relationship between urban organization and capitalism on a global scale. Applying location theory to corporate administrative structure, Hymer postulates a global hierarchy of cities corresponding to the hierarchy of the MNC. At Level 1, the *global cities* concentrate the decision-making functions of MNCs. At Level 2, the *peripheral coordinating cities* are the regional cities for MNC field offices serving as centers of coordination and control over local operations. They are fewer in number, larger, and contain more social control activities such as communications and information processing. At Level 3 are *production and marketing cities* where MNCs locate specialized branch plants according to the pull of labor, markets, and raw materials; cities of this level proliferate across the globe. In

Hymer's model, therefore, the world system of cities is a vast net cast by multinationals. A city's wealth and occupation structure; the education level of its population; and its cultural sophistication, tax base, and capacity for economic development are to a large extent determined by where its economic activities are lodged in the MNC hierarchy. The element of determinism in Hymer is seen in notions of world cities and the world system/hierarchy of cities in which the roles of cities and their international and domestic linkages are determined by their position in the world economic system or new international division of labor.

The concept of world cities applies to the top end of the world urban hierarchy. MNC growth in the 1970s and 1980s brought about integrated global production and created global markets. This in turn created a need for regional centers for administration and control. The world or global cities serve as international centers for business decision making and strategy formation (Cohen 1981). "The mode of world system integration" affects "in determinate ways the economic, social, spatial and political structure of world cities and the urbanization processes to which they are subject" (Friedmann & Wolff, 1982, p. 313). In particular, primacy, macrocephaly, and the development of world cities are at least in part, outcomes of the way in which nations are integrated into the world economy.

To provide planners, geographers, and urban economists a framework for studying future world economic changes and their impact on cities, Cohen (1981) explores possible future trends in the new international division of labor and urban hierarchy, with detailed analysis on growth of the world urban hierarchy of corporate headquarters and sophisticated corporate services in the industrialized world, particularly in the United States. He notes that the new division of labor has led to an industrial restructuring and a recent round of change in the U.S. urban hierarchy, leading to the emergence of global cities such as New York and San Francisco that are tied to the increased centralization of corporation-linked functions. Some medium-sized cities are threatened by firms relocating to less expensive sites, often in developing countries.

One consequence of such changes in the United States is the growth of new centers of corporate development or national development in some LDCs. A few of these LDC centers also serve as centers of corporate services and form part of thc new international financial markets, acting as specialized havens from high corporate taxes. They include Panama, Bahrain, Singapore, and Hong Kong. Although industrialization has advanced as a result of the new international division of labor in some LDCs, local populations have not benefited, giving rise to a dualistic structure of the urban economy and a situation of urban poverty. Cohen concludes that the future of the city in LDCs must be analyzed in relation to international and national developments. Certain exogenous forces will continue to exacerbate the flow of people into cities of LDCs, such as the international sourcing of raw materials and the use of land for export-related agricultural activities arising from the new international division of labor. At the same time, considerations like the need for entrepreneurial and technical skills, information from personal contacts, and the resolution of problems through extralegal procedures all contribute to the agglomeration of corporate manufacturing in major cities of LDCs.

To Friedmann (1986), world cities are "basing points" for global capitalism. They differ from each other, however, not only by their mode of integration with the global economy but also by their own historical past, national policy, and cultural background. The economic variable is the decisive one. Seven postulates illustrate how cities are integrated with the new international division of labor and social problems associated with this process:

1. The form and extent of a city's integration into the world economy and the functions assigned to it in the new international division of labor will be decisive for any structural change occurring within it.
2. Key cities in the world are used by global capital as basing points in spatial organization and articulation of production and markets. The resulting linkages make it possible to arrange world cities into a complex spatial hierarchy of core cities and semiperipheral cities.
3. The global control functions in world cities are directly reflected in the structure and dynamics of their production sectors and employment.
4. World cities are major sites for concentration and accumulation of international capital.
5. World cities are points of destination for large numbers of domestic and/or international migrants.
6. World city formation brings into force major contradictions of industrial capital, among them spatial and class polarization.
7. World city growth generates social costs at disproportionately high rates for poor workers and the world city itself.

In arguing that cities are centers for the accumulation of locally derived surplus for investment by foreign and local enterprises, Armstrong and McGee (1985, Figure 2) see relating cities to the world urban hierarchy as particularly germane to any study of the role of urban centers as development poles in the national economic growth process. To them, the hierarchy illustrates how decision making, surplus transfers, and capital accumulation are centralized at different levels even as ideology and behavior patterns, together with policy decisions in particular, are diffused to regional centers and rural areas.

Global cities such as New York, Tokyo, and London are the world's financial centers and the national headquarters of most MNCs. They concentrate on global accumulation and decision making. Continental cities are similar in role to the first-level cities, but within a more restricted ambit, such as Singapore and Hong Kong in Asia or São Paulo and Buenos Aires in Latin America. They serve as major subaccumulation centers. National cities like Caracas or Bangkok dominate their respective countries. They are the focuses for national accumulation and via MNCs the national links to global centers, interacting with and influencing regional cities and regulating the countryside through a system of intermediate centers.

World city and world urban hierarchy theorists seem to agree that the power to influence the relationship among cities in LDCs is concentrated in the core, whereas the key economic actors in core metropolises are MNC headquarters, banks, and advance corporate services. These exert force and dominance over the global flow of information, capital, goods, and services. The concepts also emphasize the gulf between the core and the periphery. The benefits of bringing into consideration international forces on the urbanization process of a LDC are obvious. But the emphasis on aggregated top-down determinism of the urban experience, vague distinctions between the levels of cities, and the neat vertical dimensions and functional characteristics specified for cities at each level have met with criticism (Meyer 1986b). These points require further study. Nonetheless, the implications for the world urban hierarchy of transnational capital flows are quite clear. So is the contention of the dependency theorists that LDCs are locked into certain functional roles and that their patterns of urban development are heavily influenced by these external flows.

Implications for Policy and Research

Two very fundamental and related points emerge from this review. First, the world's countries are increasingly linked through the internationalization of finance and production. Second, transnational capital flows and the behavior of MNCs exert significant influences on the current urbanization process in many LDCs. But disagreement on the nature and consequences of the process and lack of clarity and understanding of many details of linkages remain. For example, is transnational capital flow inherently urban biased or even big-city biased? Does it lead to the host LDC's disinvestment and deindustrialization in the long term? How significant is foreign direct investment as a catalyst for economic growth and general modernization of the LDCs? What impact has foreign direct investment had on regional output and income multipliers? What are its implications for specific labor skills? How has it assisted in technology transfer?

Another area needing additional study is MNC behavior. Their strategic planning processes, interrelating of diverse functions, choices of forms of joint venture, changing organizational structures, risk and uncertainty assessment, and sensitivity to international codes of government all affect their choices of host location and impact on local economic and urban development.

At the macrolevel, implications of changes in the world economy and advances in transport and communications for LDC urban development are now gaining greater attention. To improve our understanding of these issues will require substantially more theoretical and empirical work. As these issues combine exogenous and endogenous factors, they require a broad range of inquiry, from the microscale firm/sectoral level, to the mesolevel of region/country, to macrolevel international analysis.

In approaching issues related to transnational capital flows and urbanization of LDCs, there are, however, difficult and sometimes intractable data problems (e.g., national transnational capital flows data may be available, but they are rarely broken down by rural-urban districts, or even by regions). At the individual firm level, MNC confidentiality restrictions mean that answers to some of the basic questions raised above will be difficult to acquire. Simply documenting the spatial distribution of the monetary amount, employment, output value, and sectoral distribution of transnational capital flows without details on the behavior of individuals and corporations involved will preclude a deeper understanding of the dynamics of location and other regional impacts of transnational capital flows.

LDC governments are acutely aware that transnational capital flow is dramatically increasing and may be a positive element to integrate into their development strategies. In recent years, many have relaxed controls on these capital flows into their territories, devising new or improving on old incentive packages for attracting various forms of foreign direct investment (UNESCAP/CTC 1986). These governments are equally aware of the need to intervene in the development of their urban places (Pernia 1988). Yet, there appears to be little connection between the two types of official awareness—matching transnational capital flows policy with urban policy. The urban policy debate has so far focused on the location of industrial development, though the crux of the problem goes beyond what an industrial location policy can solve (Logan and Salih 1982). We need to recognize the divergence between the "economic space" of an increasingly international production system and the territorial space of the local community. The community or nation will need to channel the international flow to its spatial, economic, and social benefits.

At the moment, policies on private transnational capital inflows of most LDCs are related to incentive packages arising out of sectoral and economic considerations, with few or no locational guidelines. For some countries, there is a vague and implicit attempt to influence foreign direct investment location through the so-called infrastructural instrument, such as enterprise development zones that are often part of a country's overall dispersal policy. Such conventional locational policy would not work on this investment easily because the MNC or foreign investor is in a strong bargaining position and can threaten to take the investment elsewhere. There is a built-in tendency for market forces to counteract any attempt to divert funds from the most profitable areas (Abumere 1982), so it is not difficult to site locations of concentration of foreign direct investment contrary to the spatial dispersal policy of the host government. At this juncture, many details of the policies linking transnational capital flows to an LDC's overall urbanization policy need to be spelled out.

The long-term consequences of transnational capital flows in LDCs are also of concern. Armstrong and McGee (1985) mapped three possible scenarios for Latin America: spread of urban-led, peasant-based insurrections; return to civilian nationalist regimes pursuing a policy of nationalization and more internally oriented and directed growth; and popular opposition of continent-wide proportions to the activities of capitalism. Other speculations consider East Asia (Forbes and Thrift 1987). Taking a structuralist view that transnational capital flows generally benefit the host country, these authors spelled out the fear of keen competition among LDCs for externally sourced capital that may weaken their bargaining position vis-à-vis MNCs. They suggested formulation of international urban policy for a broad region to enable cities to coordinate their control over international capital to provide some breathing space for them to reflect on the consequences of their open competition for foreign capital. Which, if any, of these possibilities would generally fit LDCs? Clearly, much work remains if policy-relevant progress is to be made.

10

The Impact of Temporary Migration on Urban Places

Thailand and China as Case Studies

SIDNEY GOLDSTEIN

Migration is an important element in urbanization and modernization, although the specific interrelations among development, urbanization, and population mobility have not been fully delineated (Findley, Chapter 1, this volume; Lee 1966; Ravenstein 1885; Shaw 1975; Zelinsky 1971). Migration can relieve population pressures on overcrowded rural areas, populate thinly settled regions, achieve better equilibrium in the labor market between skills availability and needs, and provide the critical mass necessary for development of infrastructure.

Such redistribution may be central to modernization processes, but it is often also cited as causing serious problems of overurbanization, contributing to inadequate urban infrastructure, and negatively impacting rural development (UNDIESA 1986f, 72-75). Nonetheless, the lure of cities remains great,

---

AUTHOR'S NOTE: Research on which this chapter is based was supported in part by a Hewlett Foundation grant to the Population Studies and Training Center (PSTC), Brown University, and a joint Rockefeller Foundation grant to the PSTC and the Wuhan University Population Research Institute. I wrote the initial paper as a Senior Visiting Fellow at the East-West Population Institute, East-West Center. I gratefully acknowledge Alice Goldstein's extensive assistance.

and rural-urban migrants generally consider themselves better off in the city —even under poor circumstances—than in their home villages.

Many cities, too, depend on migrant labor to perform services and unskilled labor. Rural-to-urban migration brings benefits and problems, and a host of development factors help to determine the impact of internal migration in a particular context (Simmons 1984). A growing number of economists and other experts (Jones 1989; UNDIESA 1984) thus advocate that most nations are best advised to design policies and programs to ease the adjustment problems associated with migration and increase the benefits of migration. Only in some nations are policies designed to slow or redirect migration warranted.

Compounding the complexities of the role of permanent migration in urban growth and development is the simultaneous inflow of temporary migrants into urban places. A growing body of evidence indicates that temporary, largely circular movement, by migrants who intend no permanent change in residence and who often return to rural areas repeatedly, plays a key role in individual and household adjustment strategies to changing conditions at rural origin and urban destination, in the community, and in national development processes generally (Chapman and Prothero 1985; Prothero and Chapman 1985; see also Goldstein 1978; Prothero 1987). Circulation may help to adjust labor supply and demand without the social dislocation of large-scale permanent migration. It may also allow for informal-sector provision of urban services that are otherwise in short supply. Concurrently, the flow of funds and ideas that circulation engenders may become a critical component of rural modernization.

Permanent migration and circulation are important. In some situations one form of mobility may stimulate or substitute for the other. We know little about these relations because of the limited research on temporary movement. As a result, scholars and policymakers have been slow to recognize the complementarity of permanent and temporary movement or their joint role in population redistribution and urban growth. As more developing countries express concern about big city and rural-urban migration problems (UNDIESA 1988c, 204), more attention to the many forms taken by mobility and the different impacts that result may provide a basis for alternative distribution policies.

Progress in understanding the dynamics of mobility and its role in urban growth has been hampered by poor conceptualization, misleading perspectives on the issues, and data deficiencies (Findley 1982; Goldstein and Goldstein 1981). Using internal migration as an umbrella concept masks the differential impact that various forms of movement may have on the urban structure and the dynamics of change.

In early attempts to draw some generalizable conclusions from European data on migration, Ravenstein (1885) recognized the existence and possible importance of temporary movement, but neither he nor scholars who followed him tried to include such movement in their research. By the 1970s, however, wider acknowledgment of the role of temporary movement in the broader context of migration and development had occurred (see Zelinsky 1971, 1979).

Others, notably Chapman and Prothero (1985), recognize the diversity across time and space that characterizes population movement and the need to distinguish among various subtypes. A growing body of evidence indicates that a considerable proportion of all moves involve one or more return moves to the places of origin, and that "circulation, far from being transitional or ephemeral is a[n] . . . enduring mode of behaviour, deeply rooted in a great variety of cultures and found at all stages of socioeconomic change" (p. 6).

Chapman and Prothero provide valuable insights into what is known to date and point to what remains to be tested more rigorously. In contrast to the views of scholars who argue that wage labor circulation stems from the uneven spread of capitalism in peasant and tribal societies (e.g., Standing 1985), Chapman and Prothero maintain that circulation in the developing world is indigenous and that externally generated changes (e.g., services for political administration, religion, schooling, health, and commerce) have reenforced customary circuits of mobility and added new ones (cf. Goldstein 1988). Circulation in the past has occurred at village, community, family, and individual levels. It stemmed from a wide range of factors, such as ecological conditions, customary life-styles, trade, and political and social factors.

Basically, circulation involves a territorial separation of obligations, activities, and goods. In the Third World, it reflects the security associated with place of origin and a spatial diffusion of opportunities and associated risks that lets rural households maximize family welfare while minimizing risks. The social structure is thus bi- or multilocal and the varied destinations represent a sociospatial extension of the home community.

Evidence suggests continuing commitments to origin through the flow of communications, goods, and cash remittances and through extended visits and eventual return migration. Yet Chapman and Prothero (1985) recognize that definitive answers are needed as to whether temporary migration (circulation) will diminish or even cease as higher levels of urbanization are reached and a different or better balance is achieved between the centripetal and centrifugal forces that now account for such mobility.

If we accept Chapman and Prothero's proposition that recent or current forms of movement, including circulation, represent extensions of earlier modes of mobility, then externally generated changes have been able to take advantage of previous varieties of mobility and the flexibility of local structures to meet new needs for the territorial fluidity of wage labor. Yet temporary movement raises new issues for the destinations. If individuals are not fully committed to their places of destination, but regard themselves only as sojourners, their demands on the temporary destination and their patterns of behavior may be different from those of permanent settlers with respect to kinds of housing, need for amenities and services, friendship and affiliation patterns, and use of savings (Nelson 1976).

Full evaluation of migration and its impact on city populations requires separate attention to movement resulting in either permanent change in residence or temporary mobility. Both may affect the size, structure, and dynamics of cities, but quite differently. We must first recognize that the de facto number of city dwellers may be very different from the number official census and registration statistics indicate because a substantial proportion of temporary migrants are often not enumerated as residents. The actual numbers functioning in big cities, and especially in the labor force, may thus be as much as 10-20 percent greater than the official resident population.

The implications of temporary mobility go well beyond mere population size. Temporary and permanent migrants differ from each other in their duration of residence and in a host of socioeconomic characteristics. Both groups also differ from nonmigrants. Thus, questions must also be raised about the effect of temporary migration on the socioeconomic and demographic structures of the urban communities, the temporary migrants' welfare, their links to their home communities, the burdens they place on the infrastructure at their destinations and how this differs in quantity and type from that imposed by permanent in-migrants (Rondinelli 1989), and the impact of bilocal residence on the quality of life in the communities of destination and origin.

The limited data available for assessing the comparative roles of permanent and temporary migration in the urbanization process in particular countries, as well as the even more serious problems that arise for comparative cross-national research, argue strongly for restricting an evaluation of the impact of migration on urban places to those data sets and countries that allow the fullest assessment (for an evaluation of different data sources, see Findley 1982; Goldstein and Goldstein 1981). Here I draw heavily on data from Thailand and China because I have worked extensively in both countries and have either directly studied migration in them or have influenced—sometimes not as effectively as I hoped—the research of others. Data from these nations illustrate the demographic effects of various forms of migration, especially temporary movement, on urban places.

Development and Urbanization in Thailand and China

China and Thailand are among many developing countries where population redistribution has assumed great importance in demographic change and development. Both have experimented with various policies to control migration to their big cities or divert movement elsewhere. Temporary population movement seems for both to be an important mechanism for coping with rural pressures, urban constraints, and urban labor force and service needs. Thailand and China differ greatly in size, development, political systems, and the ability to control migration and urban growth. Yet each has experienced rapid changes in development since 1960, and in the early 1980s had comparatively low urbanization levels (about 20 percent).

Economic Development in Thailand

Compared to many other developing countries, Thailand has experienced quite rapid socioeconomic change since 1960. Its per capita GNP grew faster than that of all but 11 other developing countries (UNESCAP, Population Division 1984). Although the majority of Thailand's population has participated in and benefited from development, differentials persist among the country's regions and between rural and urban populations. The country remains largely agricultural (in 1985, 80 percent of the population was estimated to be living in nonmunicipal areas), but a substantial shift from self-employment in agriculture to employment outside the primary sector provides continuing stimuli for permanent migration and temporary movement.

The Thai government, unlike its Chinese counterpart, does not directly control migration. Its policy has been to decelerate the trend in rural-to-urban movement through reducing imbalances in economic development and accelerating development of regional urban centers as intermediaries between provincial cities and villages. Regional urban centers and small towns were seen as requiring support in the form of physical and social infrastructure and monetary investment (Vichit-Vadakan 1984). Nonetheless, Bangkok, the world's most primate city (Sternstein 1984), continues to attract migrants from rural areas, smaller urban centers, and even from cities designated as regional development centers (Vichit-Vadakan 1984, 490).

Population Policy and Development in China

Chinese policymakers also view population distribution as needing planning and have tried to control it directly as part of broader central planning and development efforts (for more detailed discussion of China's migration policy, see Goldstein 1990). China's political system allows a high degree of control, so its experience is an interesting contrast to Thailand's. The Chinese model may provide insights for avoiding some of the negative consequences of rapid urban growth while simultaneously realizing the benefits of urbanization, absorbing surplus rural labor, and achieving economic development.

As in Thailand and elsewhere, substantial differences characterize the quality of life between urban and rural places, despite ostensible efforts to reduce them (Johnson 1988). Urban places have higher incomes; better housing, sanitation, and educational opportunities; subsidized food; more varied entertainment; and greater availability of consumer goods. Urban job security has been higher, and a wide range of benefits has been associated with employment in state-owned enterprises. All have been powerful magnets for peasants whose annual incomes depended on the vagaries of weather and, until the 1980s, on income distribution decisions by the collective leadership of the communes. Some of these conditions are changing as a result of economic liberalization in China. Nonetheless, the attraction of urban places for peasants remains quite strong (cf. Zhu and Wang 1985).

Evidence from the mid-1987 National Survey of the Chinese Population (China. State Statistical Bureau 1988) clearly points to cityward migration as the most popular direction. Three-fourths of all movers were migrants to urban places compared to only one-fourth who moved to rural locations. By contrast, two-thirds of all migrants had rural origins. Thus, urban places experienced a net gain of 12.9 million persons, of whom 5.7 million had settled in cities and 7.2 million in towns. Clearly, substantial population movement has occurred as a result of the diversification of urban and rural economies.

Yet this mobility evidently represents only the tip of the iceberg. Estimates indicate that by the late 1980s China had more than 50 million transients in its population (Shanghai Population Information Centre 1989), a disproportionate number of them in the large cities. The 23 cities with 1 million or more population encompassed about 10 million transients. Much of this temporary migration is a result of the changing economy, as indicated by the finding that two-thirds of the transients are peddlers, individual crafts workers, or construction workers. Most of the rest are business travelers or visitors.

In turning next to a comparative evaluation of cityward migration in Thailand and China, I focus first on Bangkok as Thailand's preeminent city, then on several of China's giant cities.

Migration Patterns: Bangkok

Lifetime Migrants

In 1960, when the Thai census first asked a migration question, 22.8 percent of Bangkok's population reported having been born outside Bangkok's provinces of Phranakhon and Thonburi. From 1960 to 1970, Bangkok grew from 2.1 to 3.1 million, with 27 percent of its population born elsewhere in Thailand.[1] In 1980, Bangkok's percentage of lifetime migrants was virtually the same as in 1970, reflecting the combined effects of the mortality and out-migration of earlier in-migrants and the increase in the number of Bangkok-born persons.

5-Year Migrants

Data on 5-year provincial migration allow more precise determination of interdecade changes. According to the 1960 census, 7.3 percent of the capital's population had moved there since 1955. Reflecting 1960s dynamism, by 1970 the number of 5-year migrants in the capital had doubled, forming 11 percent of Bangkok's population. By 1980 the pattern had changed, reflecting in part a diminution in the 1970s of the dramatic pace of national economic and social development in the 1960s. Despite an increase in the absolute number of recent migrants to

Bangkok from 299,000 in 1965-1970 to 341,000 in 1975-1980, their proportion in the population declined to 8.0 percent.

The extensive turnover in Bangkok's population is evidenced by the heavy ratio of out- to in-migrants. Whereas 341,000 persons moved to the capital between 1975 and 1980, 170,400 moved out during the same interval. Whether the out-migrants consisted disproportionately of earlier in-migrants cannot be ascertained, but some insights can be gained from assessing repeat migration.

The major importance of migration in the growth of Bangkok's labor force is provided by estimates that after adjustment for conceptual changes, about 60 percent of its 1975-1980 growth was attributable to net migration (UNESCAP 1988). Migration was responsible especially for the increase in service workers and played a key role in increases in production and transport workers. The ESCAP study stressed that high rates of temporary migration have probably helped to alleviate many of the social and economic costs associated with rapid growth, because the burden of financing education, housing, and health care for a hefty portion of Bangkok's work force is borne by the temporary migrants' places of origin.

Repeat and Circular Mobility

Cross-tabulations of lifetime by 5-year migration status suggest that the great majority of Bangkok migrants came directly from their birth province and that most arrived before 1975. Of those arriving since 1975, one in four had moved more than once, with 8.8 percent being return migrants to Bangkok and 13.4 percent migrants from places other than their province of birth. As many as one-quarter of Bangkok's migrants had lived in three or more different urban places; repeated residence in Bangkok was common among those living in other urban Thai places, where a third had lived in Bangkok at least once (Goldstein and Goldstein 1979).

Although not specific to Bangkok, comparison of lifetime and recent migration suggests a very substantial increase in the extent of repeat and return migration in Thailand. According to the 1970 census, 28 percent of those who moved between 1965 and 1970 had made at least one previous move. By 1980, 39 percent of the recent migrants (1975-1980) had done so. Moreover, one-fifth of the more recent movement was attributable to returns to the province of birth, compared to only 10 percent in 1965-1970. A major part of the repeat and return movement may reflect the inability of urban areas, especially Bangkok, to provide employment for in-migrants and city youths as they reach working age. Faced with unemployment or underemployment, migrants to urban centers may opt to return to their place of origin or try still another location.

Further, by the mid-1970s, government development efforts in rural areas may have been successful enough to lure migrants to their place of origin. Improved transportation networks may also make circulation between village and city more feasible. Periodic movement, as economic opportunities arise elsewhere, may be quite attractive to risk-averse individuals and families.

Since 1974, annual migration surveys in the Bangkok Metropolitan Area provide better insights on short-term migration to the capital than does the census. The 1974 survey estimated that 70,600 persons (2 percent of Bangkok's population) had in-migrated between November 1972 and June 1974 (Thailand. National Statistical Office 1976). A slight majority were males, but this varied by age, with many more women (84:100) in the 10-19 group and many more men (132:100) in the 20-39 range. Migrants were heavily concentrated in the younger ages, with 75 percent aged 10-29 and 12 percent under 10. Over two-thirds were single, and most moved for economic reasons (54 percent to find work). Almost three-fourths were farmers before moving to Bangkok, but once in the capital most men were engaged as craftsmen, production and process workers, and laborers, whereas most women became service and recreation workers.

A substantial part of the movement seems to be seasonal or of tentative duration: About one-third of those whose move was economically motivated were looking for work during the agricultural slack season. Only 5 percent of in-migrants during a 2-year period expected a stay of less than a year, and only 23 percent expected to remain over a year, including 21 percent who planned to remain in Bangkok "forever." Significantly, three-fourths could not state a specific duration. For most migrants, onward mobility or a return to place of origin appears to be an open possibility.

Relatively little increase occurred in the number of in-migrants to Bangkok between 1974 and 1985, although its population increased by about 50 percent. Thus, the 84,060 recent in-migrants made up only 1.4 percent of the 1985 population. Although in-migration declined, in-migrant characteristics in 1985 generally remained quite similar to 1974—migrants remained heavily concentrated in the younger ages, most were single, and economic motives drove most of the moves. The major change in composition from 1974 to 1985 was the much larger percentage of women who migrated temporarily. The 1985 survey found only 64 males per 100 females among the migrants; even in the 20-39 age group the shift to females was dramatic, as a sex ratio of only 77 males per 100 females demonstrates (cf. Smith, Khoo, and Go 1984).

Again, although the largest number by far, 78 percent, were farmers before the move, few remained in agriculture after moving. By the 1985 survey, over two-thirds of male farmers worked as craftsmen, production workers, and laborers, and another 12 percent were in transport and communications. A majority (58 percent) of women who had been farmers were employed as service workers, and another third were craftswomen, production workers, and laborers. Thus, a high degree of occupational mobility was associated with the change in residence.

In 1985, too, most in-migrants seemed very uncertain about the permanency of their move to Bangkok. Only 6 percent definitely foresaw a stay of less than a year; 23 percent expected to remain over a year; 9 percent anticipated staying "forever." That 62 percent could not fix a firm duration for their residence in Bangkok again suggests that for most migrants returning to origin or going to another location remains possible. Yet, compared to 1974, the 1985 data point to some increase in the proportion expecting to remain in the capital longer than a year.

Among economically motivated migrants, a much higher percentage in 1985 had moved to seek work during the slack season than did so in 1975 (88.7 versus 28.5 percent). This finding suggests that much of the movement is highly temporary, and probably increased substantially in the 1980s; data show only 12 percent had registered in their new location. This was true of 30 percent of migrants from municipal areas but of only 8 percent of those moving from rural locations, suggesting that those from other urban locations were more likely to regard Bangkok as a permanent change of residence. Of course, rural migrants may also be less familiar with registration requirements and so less likely to register.

Most important, fewer than one-third of urban and rural migrants to Bangkok changed their registration; a high proportion could not specify how long they expected to remain in Bangkok, 55 percent had moved alone, and a substantial proportion were obviously seasonal laborers. These findings strongly support the conclusion that a considerable part of the reported movement to Bangkok is temporary and likely to lead to future out-migration. Over half of those who expected to remain in Bangkok less than a year indicated the northeast as their likely destination (this region also accounted for the single largest source of in-migrants), suggesting that for many potential out-migrants, movement takes the form of return migration.

Further supporting the importance of temporary mobility is a study by Chamratrithirong, Archavanitkul, and Kanungsukkasem (1979). Efforts to recontact a sample of migrants identified in a survey of Bangkok taken 7 months earlier revealed that just over one-fifth of the men and one-third of the women who had arrived in Bangkok within 2 years preceding the initial survey had already returned to their communities of origin. Oddly, migrants with jobs at the time of the first interview were more likely to have left than were those looking for work or not working. Half of these women and 43 percent of the men had not registered their Bangkok residence because they considered their stay temporary.

Together, these studies strongly suggest that the more permanent movement identified by census data represents only a small portion of total movement to and from Bangkok (cf. Goldstein 1978). Without attention to circulation and other forms of return and repeat migration, the relation between population movement and big-city growth and development cannot be fairly assessed. Many forms of movement are used to allow individuals to take advantage of the economic and social opportunities available in Bangkok. At the same time, the growing network of other urban locations, including small places, provides an increasing number of destination alternatives for different types of movers.

Migration Patterns: Giant Cities in China

Until the late 1980s, a broad assessment of migration in the People's Republic of China was limited by a paucity of data, especially at the national level (Goldstein and Goldstein 1987). In part, this situation reflected the decentralized nature of the statistical record system, in part the widely held belief that local registry statistics provide adequate information because migration was largely controlled through the registration system. Thus, the 1982 census of China did not include any direct migration questions.

Because of this lack of census data on migration and the limited migration statistics available from the registers, especially as related to temporary migration, sample surveys provide the most promising information on migration in China. Reflecting the growing interest of policymakers, planners, and scholars in migration, an increasing number of such surveys have been initiated (Goldstein and Goldstein 1987). My discussion of migration patterns and migrant types draws on a variety of studies. Only a brief description of each data source is given. Readers interested in more detail are encouraged to refer to the fuller references.

An Omnibus Survey of Shanghai (Goldstein and Goldstein 1990), conducted in 1986, collected information on permanent in-migration to the city from 1,000 individuals. Although definitions used here conform to Chinese criteria (*permanent in-migrants* are only those persons who have legally obtained registration in Shanghai) and are therefore not comparable to the Thai statistics, the Omnibus Survey does allow disaggregation of the in-migrants into lifetime and 5-year categories as the Thai census does.

The Chinese Academy of Social Sciences (CASS 1988) National Migration Survey was conducted in 1986 in 74 cities and towns in 16 provinces and included nonmigrants, permanent in-migrants, and temporary migrants living in the households sampled. *Temporary migrants* were defined as those who had moved to their household of residence without an official change of registration and had been living there less than a year. Anyone moving to the current residence over a year before the survey was considered a permanent in-migrant regardless of registration status. No opportunity exists for analysis of this data set beyond data available in published tables.

The Survey of Migration, Fertility, and Economic Change in Hubei Province was conducted in 1988 and covered rural and urban places (Goldstein and Gu 1991). The sample included people living in households, free markets, and construction sites. The survey allows identification of nonmigrants, permanent in-migrants, temporary migrants, and commuters. Permanent in-migrants are people who changed their registration in connection with their move to current residence. Temporary migrants and commuters are registered at a place other than the interview site, but the former consider the interview site their current residence; commuters indicate their residence is at place of origin.

The 1984 Shanghai Floating Population Study (Goldstein, Goldstein, and Guo 1991) and the 1986 Beijing Survey of Temporary Migrants (Goldstein and Guo 1991) are restricted to central city districts in each municipality. Each survey sought information on temporary migrants in households, free markets, construction sites, and other temporary quarters. The former collected detailed information only on temporary migrants living in households. The latter also concentrated on those in households, but collected some individual characteristics for migrants in hotels, hospitals, and construction sites. Neither survey sought information on nonmigrants or permanent in-migrants.

Because I am concerned with the impact of temporary migration on giant cities, my discussion of migrant characteristics and migration streams is restricted to data from the CASS Survey that focus on cities with 1 million or more population and to Wuhan from the Hubei Survey. The Shanghai and Beijing surveys are used mainly for data on migrants living in households.

Types of Migration to Giant Cities

Largely reflecting the development of industry in Shanghai during the 1950s and 1960s, one-third of the city's 1986 population reported having been born elsewhere in China. By contrast, only 2.8 percent of Shanghai's population reported moving into the city as registered residents between 1981 and 1986.

Among 1986 residents, 67 percent were born in Shanghai and had been living there 5 years earlier in 1981. Judged by this measure, a considerable majority of the city's population was residentially stable, quite similar to the 64 percent of Bangkok's 1980 adult population who were born in the capital and living there in 1975. Only 1 percent were born in Shanghai, had moved out, then returned after 1981. Of the 32 percent born outside the city, 95 percent had moved to Shanghai before 1981, a level well below Bangkok's.

In Shanghai as in other Third World cities, a considerable portion of the population was born elsewhere. Shanghai has only a small percentage of recent permanent in-migrants, however. Reflecting the effectiveness of the migration control policy, only 3 percent of all of Shanghai's registered population age 15 and over had moved to the city since 1981, and half of them were return migrants. Yet this finding provides a misleading picture of migration to Shanghai and, by extension, to China's giant cities in general. Permanent in-migration is only one segment of population mobility; temporary movement has become an especially important part of total mobility to Chinese cities.

Economic reforms introduced in rural China since 1979 have exacerbated the surplus of rural labor even as economic opportunities for rural residents increase in urban places. Because of restrictions on permanent migration, few rural persons can change their registration to giant cities. They can, however, move to these cities temporarily by maintaining their rural registration. Rural people go to cities to market products, engage in construction work, and provide a wide range of services.

The 1984 Shanghai Floating Population Study estimated 586,000 temporary migrants in the city,[2] 58 percent living in households, 22 percent in hotels. Temporary migrants living on boats or around the Shanghai periphery constituted 7 percent, and others, including construction workers and peddlers in free markets, 13 percent. Since central Shanghai's 1984 registered population numbered 6.32 million, temporary migrants increased that number by 9 percent, attesting to their importance in China's urban population growth.

The 1986 CASS Survey also evaluated the volume of temporary movement in relation to permanent in-migration, but because the CASS Survey interviewed only persons in households and used a restricted definition of *temporary migrant,* these data underestimate the level of temporary migration.

Of the 44,723 persons sampled in the giant cities, two-thirds were nonmigrants. Among the 15,092 permanent in-migrants, only 1,222 had arrived at their destinations in the 2 years preceding the survey (1985-1986). This number is below the 1,519 who entered the giant cities as temporary migrants in 1986 alone. Compared to the large volume of temporary migration, annual permanent in-migration is more modest and apparently being successfully controlled under China's migration policies.

The Impact of Temporary Migration

Beijing shows the impact of temporary migration on city size. At the end of 1987, the city had a registered population of 6.71 million. In 1988 it also had a floating population (people without permanent registration in the city) of 1.15 million, including 730,000 engaged in manual work and peddling (*China Daily* 1988c). It has been estimated that the total average size of the floating population in China's 23 largest cities numbered 10 million.

Temporary workers contribute to urban development by offsetting labor shortages in areas such as construction and services. They also fill needs that local residents are not qualified or willing to meet, such as making furniture; repairing shoes, watches, and umbrellas; and cleaning streets. A number of enterprises prefer rural laborers because they accept even the hardest work at very low pay (*China Daily* 1989a). This has the effect, however, of reinforcing the unemployment rate of the local population. Ironically, even though most cities have many unemployed youths, it is still difficult to fill many positions that involve undesirable work conditions.

High urban growth rates already cause considerable concern about strains on urban infrastructure. The large numbers of temporary migrants in cities exacerbate the problems of providing adequate jobs, transport, energy, water, housing, goods, and materials. It has been estimated for Shanghai, for example, that each new member of the floating population costs an additional 600 yuan of investment in nonproductive service facilities (*China Daily* 1989c). Underwriting the costs of coping with the more than 1 million temporary migrants in Beijing seriously interferes with efforts to develop satellite areas as a way to divert people from the dense core areas to the outskirts (*China Daily* 1990).

As a result of exacerbated pressures, some city governments, such as Beijing's, are developing plans to curtail by as much as half this large-scale movement into the city and to require licensing of temporary workers. They hope to reduce the negative impact temporary migrants are seen as having on the city's overall welfare. This situation points to the need to find ways to articulate large and growing rural surplus labor with nonagricultural activities in towns and villages and also to use skilled city surplus labor productively in towns and villages to create nonagricultural jobs (*China Daily* 1988a).

Some migrants move back to rural areas under pressure from urban governments (see, e.g., *China Daily* 1991b), especially after slowdowns in industrial growth and cutbacks in capital construction. Others return home because they feel they can make a better income on the land than in industry, construction, or commerce. Despite the exodus of many peasants to rural areas, however, the trend of labor's moving to urban places persists; only the rate has declined. Because many are unsuccessful in finding legitimate means to make a living, the street population of China's major cities has been growing in the last few years; and many street people, usually migrants, resort to illegal activities to survive (*China Daily* 1988b). With no easy access to housing and sanitary facilities, such migrants may become health threats as carriers of contagious diseases. Many sleep at railroad stations, in underground sewers, under highway overpasses, in parks, and at construction sites. Some cope by becoming beggars (*Beijing Review* staff, 1988). Efforts in Beijing to return such "vagrants" to their home communities have cost the municipality 1.5 million yuan ($400,000) a year (*China Daily* 1988b). Nationwide, it has been reported to cost as much as the equivalent of $12-15 million.

Among reported by-products of greater population mobility in China is an inflated birth rate. A substantial number of temporary migrants circumvent the family planning policy by having babies while away from the community in which they are officially registered. As most are also not registered at their destination, and so not under control of a family planning unit, they have greater freedom to deviate from established family planning regulations. Moreover, because many temporary migrants enjoy better than average incomes, they do not feel threatened by possible economic penalties if their violation of fertility policies comes to the attention of authorities. Incorporating such migrants into the control system presents a major challenge to China's efforts to achieve and maintain low fertility when increased mobility has become an inherent part of the scene (*China Daily* 1989b; 1991a).

Temporary Migration: Selected Cross-National Perspectives

A very high proportion of developing countries, including China and Thailand, regard their current patterns of population distribution as unacceptable and have adopted policies intended to correct the situation (UNDIESA 1988c). Particular concern focuses on problems associated with big cities, their role in the urban hierarchy and in the development process, and the still-growing rural population. The latter will have to have adequate chances for livelihood in rural areas or smaller urban places if migration to big cities is to be avoided. Redirecting population movement and rural development are seen as ways to achieve these goals, although few countries have so far been successful in such efforts. Reliance on other forms of mobility, such as circular migration and commuting, represent other alternatives, although even less is known about these processes and how effectively they may contribute to solving problems associated with urban growth and population redistribution.

Thailand and China provide good examples of nations struggling with these issues, sometimes in similar ways and sometimes quite differently. In Thailand, a wide range of policies has directly or indirectly favored urban places, especially Bangkok. These have resulted in heavy rural-urban migration and the unparalleled primacy of the capital. By the mid-1970s, however, the situation had changed considerably. Government investments in regional urban centers encouraged their growth, and although Bangkok and its surrounding provinces remained attractive destinations for migrants, other cities also drew movers.

Thus, even without direct policies to control population movement, through a variety of development policies, Thailand has indirectly affected migration streams (Lim and Schilling 1980). That Bangkok's primacy continues and that rural/urban inequalities remain great testify to the still-limited success of these policies in meeting national development goals. Further, the rural/urban disparities have evidently motivated many rural residents to turn to circular migration as a way of coping with the situation and minimizing risk.

China's attempts at strict control of permanent migration and simultaneous development, especially in giant cities like Shanghai, Beijing, and Wuhan, are a striking contrast to Thailand's. Efforts to limit urban growth by directing and controlling people's movement in the face of growing rural labor surpluses have led to heavy temporary movement and commuting and to the development of small towns as alternative urban destinations. China's experience here may prove valuable for other developing countries, although the communist government's ability to institute policies more rapidly and forcefully than can governments in other nations undoubtedly creates an atypical situation.

The debate in China on the future direction of its urbanization recognizes that emphasizing the development of small cities and towns is not without its own set of problems, that this emphasis may not be adequate in the future, and that large cities must also play a central role. The designation of port cities and key-point cities, with preferential access to resources, is indicative of the trend (see Kirkby 1985, 201-248). In all such discussions, the role of permanent versus temporary migration must be considered.

Overall, temporary movement in China has come to complement and often substitute for permanent rural-to-urban relocation. Given more efficient labor use introduced by improved technology and more effective organization, and with higher motivation for profits, such movement allows and stimulates more peasants to seek the benefits of urban employment through migration. With permanent change in residence still largely impossible, temporary movement provides an alternative for many.

In China and elsewhere, circulation may also help to adjust labor supply and demand on a regional or even national basis without the social dislocation that large-scale permanent migration entails or the added strain on cities that the influx of numerous permanent in-migrants creates (Standing 1985). Temporary migration can be instrumental in redistributing labor to meet needs for skills that are not locally available and to foster informal-sector provision of urban services that are otherwise lacking. Because such movement does not involve as much official intervention, it can also respond much more quickly to labor market needs than can the more controlled permanent migration. Moreover, it allows a substantial number of individuals to engage in noneconomic mobility and thereby develop informal urban-rural links that can become an integral part of the modernization process. Concurrently, the flow of funds and ideas that circulation engenders may become a critical component of rural modernization (Chapman and Prothero 1985).

A number of studies have documented that both the nature and meaning of cityward migration vary greatly from country to country (Prothero and Chapman 1985; Standing 1985). In Latin America, the great majority of migrants leave the countryside permanently. Some may return for visits; few return to resume residence. By contrast, Africa is regarded as a region where many migrants consider their moves as temporary, either for a few years or for the period of the mover's working life. Close ties are maintained to the village of origin, and a high proportion of the migrants eventually return there. For many Asian countries, stability of the rural population has often been cited; where relevant data have been obtained, the evidence for Asia generally and for Southeast Asia in particular indicates that general patterns of population movement are similar to those in Africa.

Much of the total movement and a major portion of the urban population are hidden from researchers who fail to ask the right questions, to use the most appropriate political or geographic units of measurement, and to establish the correct overall research designs. Quite consistently, measured levels of mobility rise as the size of the units for analysis is reduced, and the amount of movement identified increases as the chance to record or observe short-term movements is enhanced. The extent of circulation or return migration is far greater than censuses reveal.

Evidence from Thailand and China discussed in some detail here and the findings of a growing number of other studies lend strong support to the conclusion that, "Throughout the world a stream of migrants covers a spectrum from the seasonal or sporadic short-term movers seeking to supplement a meager rural income to the permanent migrant attempting to substitute one set of lifetime prospects for another" (Nelson 1976, 20). What evidently varies from country to country and region to region is not the variety of migration forms but rather the particular mix of alternatives and the exact conditions under which one or another predominates. High priority should go to research identifying the factors that explain the variation in patterns and how they relate to city growth and problems of urbanization in general.

Future Research Directions

In particular, we need to try to answer the questions: What are the specific effects of temporary migration on the number of persons living in cities and on the socioeconomic and demographic composition of the population? To what extent does temporary migration relieve the problems that cities such as Beijing, Shanghai, Jakarta, and Bangkok would face if all or most of the temporary movers were permanent migrants? To what extent does such movement exacerbate these problems? Should governments encourage temporary migration as an alternative to permanent migration? If so, what would the economic and social costs be in implementing such a policy, compared to either making greater efforts at rural development or allowing additional permanent migrants into the cities? How do development efforts, such as improved transport and education systems and mechanization of agriculture, affect levels of migration and circulation? To what extent do various forms of mobility singly or combined relieve rural pressures? How much does the population interchange between urban and rural places, and particularly circular movement, contribute to the modernization of rural areas? How much do remittances and the funds and goods temporary migrants bring back contribute to reducing rural-urban inequities? Will temporary migration greatly diminish or cease as higher levels of urbanization are reached and a different or better balance is achieved between centripetal and centrifugal forces that now account for such mobility? To what extent does international labor migration complement or substitute for the various forms of internal migration?

Answering such questions requires concerted attention to circulation and commuting in all forthcoming efforts to assess Third World population movement and redistribution. In pursuing such research, we must above all recognize that the concept of *migrant* masks "a complex array of subtypes based on the variety of forces operating to stimulate the move, on the periodicity and degree of permanent commitment related to the move, and on the location of migrants within the social structure of both sending and recipient units" (Abu-Lughod 1975, 202). Rural-urban mobility needs to be conceptualized so that all movers, not just selective subsets, can be distinguished from stayers, and so that meaningful distinctions can be made among types of movers judged particularly by degree of commitment to their home place and the place of destination. Adequate attention to individual movers, the traits and conditions of their households, and their places of origin and destination should let us pinpoint the conditions leading to the decision to stay or move.

For those who move, it should also be possible to determine what individual and contextual factors account for the type of move chosen—commuting, circulation, or permanent migration—and why and how the specific destination of the move was chosen in lieu of alternative destinations. We should also be able to determine whether adjustment varies by type of move and how decisions are made to change from one form of movement to another. For all types of moves, it is necessary to assess as fully as possible the nature of the interaction with place of origin and destination as well as the effect of that interaction on stayers and on the quality of life at origin and destination.

The experiences of Thailand, China, and other developing countries strongly indicate that many people are neither exclusively rural nor exclusively urban: The dichotomies we have relied on so heavily are artifacts of our taxonomy rather than a reflection of reality. Many persons, and the evidence suggests a growing portion of the population, may have bilocal or even multilocal residence patterns. The interests of these individual movers as well as of the rural and urban communities of which they are a part and to which they make important contributions can best be served by (a) data systems that identify the temporary migrants who function in giant cities and other urban locations while still counting them in their home communities and (b) policies that take account of the needs of the resident populations in both locations as well as of those who move between. To be successful, such policies require that we recognize and give equal attention to all forms of population movement, including temporary movement, in our theoretical formulations, in our research, and in our efforts to develop effective policies.

Temporary movement may well provide an important mechanism for coping with problems of big cities, overurbanization, and rural development. In the process, it may help to reduce inequalities between rural and urban places and lead to a higher quality of urban and rural life. At a minimum, recognizing the extensive amount of such movement and its characteristics should facilitate better planning for services in giant cities and thus contribute to the welfare of the migrant and resident populations.

Notes

1. For a fuller discussion of migration in Thailand as documented by the censuses of 1960, 1970, and 1980, see Goldstein and Goldstein 1986.

2. Reflecting somewhat wider coverage and economic changes, the volume of temporary migration identified in succeeding surveys exceeded 1 million (Zheng et al. 1985a, 1985b).

11

Urban Aspects of Labor Migration

A Review of Exit Countries

FRANCISCO ALBA

People's movement across national borders for employment is a centuries-old process. It gained additional momentum after World War II with the assumption that economic growth should not be restrained by insufficient labor (Böhning 1979; Kindleberger 1967; OECD 1979). What is new about contemporary international labor migration is that it is increasingly viewed and conceptualized as evidence of an economically integrated world (Donges 1987), an *exchange* between countries in the context of other exchanges: goods, capital, technology, ideas, and even values.

During the economic boom following World War II, international labor exchanges were considered advantageous for all involved, bringing the world an overall significant rise in output, income, and employment. Labor-importing countries would supplement their work force to sustain growth and increase output—goals otherwise constrained for lack of labor. Labor-exporting countries would benefit from an easing of employment pressures, the return of a more skilled labor force, and the inflow of migrants' remittances. The betterment of the migrants and their families was also taken for granted.

The ideal labor migration (based on the West European "guest worker" migration of the 1950s and 1960s) is described as a system of labor allocation between countries that implies the "rotation" of its members, or temporariness of visits, are salient characteristics (García y Griego

1983; OECD 1979, 16; Vialet and McClure 1980; Wilson 1976).[1] Labor migration seemed to have the advanced, industrial countries as main destinations, to the point that a certain south-north direction was linked to this movement. A closer view of the phenomenon obliges us to qualify both temporariness and directionality.

As labor flows have matured, there has been some conversion from temporary moves to permanent residence abroad.[2] Similarly, labor flows are not limited to movements toward the world's core industrial areas; in a very real sense labor movements take place from less to more developed regions. Labor migration is certainly global, but composed of particular streams and flows that form distinctive regional labor markets or migratory systems. Today most countries either send or receive migrant workers. For example, important labor flows exist among Latin American countries with the same country sending and receiving labor simultaneously (Balán 1985).

By the late 1960s and early 1970s, views on the nature of international labor migration changed markedly as the consequences of a mature labor migration phenomenon began to unravel and more became known about its complexity. In industrialized countries, changes in immigration policy attest to this changing perspective. Labor-supplying countries also began to reconsider the pros and cons of emigration, and the frustrations of development became at times associated with the drain of human resources through emigration.

Despite some concerns about labor migration, it continues unabated. Böhning (1984) estimated the stock of migrant workers in 1980 as 20-21 million worldwide; although this is probably a low estimate, higher estimates of the global stock of migrant workers remain speculative.[3] International migration trends in the late 1980s and early 1990s have not been clearly defined. A less dynamic environment for foreign labor seems to be widely accepted as a fair characterization of current affairs,[4] but most scholars tend to agree that temporary labor migration seems likely to continue for the foreseeable future (Salt 1989). Certainly with current demographic and economic conditions, pressures to emigrate will mount (Golini and Bonifazi 1987).

Although a number of developing countries are already predominantly urban, and many others will reach that stage in the near future (UNDIESA 1987d), the links between urbanization and labor emigration have received scant attention.[5] This topic does not seem to command separate treatment, although there are some exceptions—the contribution by Seccombe and Findlay (1989) on the Middle East and one by Dandler and Medeiros (1988) on Latin America.

In this chapter I identify the major issues in the literature on labor migration and urbanization. I contend that the current revisionist position on the pros and cons of international labor migration needs to incorporate the urban dimension within its framework. The relevance of this exercise for policy is self-evident given the growing importance—in absolute and relative sizes—of urban populations and economies in exit countries. I provide a general overview of the role assigned to urbanization in migration and of the urban effects of the processes of migration and return migration in labor-supplying countries. I then consider the urban dimension within the debate on emigration determinants, examining various ways urban processes in sending countries are affected by emigration, concentrating on demographics, labor market effects, and remittances.

Urbanization and Labor Out-Migration

Informed, conventional thinking on international migration is based on the central theory that economic disparities between countries produce migration. Econometric exercises emphasize differentials between receiving and sending countries within a short-term orientation, relying on individual traits and associated structural factors to explain migration flows. Certain factors "push" people from source regions—demographic pressure, unemployment, low wages; others "pull" them to destination regions—high wages, employment prospects, amenities.

The push-pull framework has proved convenient for modeling purposes (see Williamson 1988), but many scholars dispute whether factors such as population pressures, unemployment, and wage differentials are the sole determinants of migration and add social and political factors to supplement the economic ones. For international migration, particular attention must focus on the role of the state and other political factors and on social factors in establishing social networks. Because international migration involves crossing national borders, many authors consider the state the major element in analyzing this issue (Böhning 1984).[6]

Against the conventional explanation of migration that assumes high emigration rates are associated with low development levels, an emerging perspective sees migration as a process evolving from an established network of relations between sending and receiving societies. The underlying conditions that fuel migration result from the integrated patterns of economic development, not from independent or separate paths of development (Portes and Walton 1981).[7] Demand (economic) forces start the labor move that later develops, driven by its own dynamics. The importance of social networks becomes evident when already-mature flows are examined. Family, friendship, and community networks are the most frequently analyzed and have come to be considered central components in the formation and evolution of migration systems. "They mediate between individual actors and larger structural forces [and] they explain the continuation of migration long after the original impetus for migration has ended" (Boyd 1989, 661). This seems true for flows as diverse as those taking place within Africa (Adeokun 1991) or between Mexico and the United States (Massey et al. 1987).[8]

Portes and Bach (1985, 111-112) offer a comprehensive formulation of the forces at work in international migration; they conceived of three sets of "stratifying relationships that organize or structure migrant flows." Set 1 includes economic linkages: capital employed, trade relations, cultural and consumption expectations. Set 2 consists of the diverse political relationships whose influence is most clearly felt through state migration policies. Set 3 covers the many social ties binding the migrant to family, kin, and friends.

Where does the urbanization process fit into these relationships? To date there is little work specifically relating urbanization processes to international migration. Migration, including international labor migration, is usually viewed as rising out of backwardness, stagnation, and rural conditions. In the literature, the emphasis on rural considerations is overwhelming. Many scholars note that international migration tends to accelerate during the transition most countries experience from a heavily agricultural economy to one with a greater dependence on urban economic activities, a transition that usually causes a profound breakdown of rural societies (Arizpe 1983; Massey 1988).

This heavy emphasis on rural conditions is understandable given that until recently most developing societies were still chiefly rural and agrarian. For many migrants the move is an escape from an environment of agricultural subsistence, though urbanization has become a dominant feature in many exit countries. This structural transformation of sending countries is reflected in the changing origin and skills of migrants.

Changing Origin of Migrant Populations

Although many scholars perceive the rural population as the "obvious" source of international migration, others argue differently: "Conventional wisdom has it that international migration begins from the cities" (Chaney 1985, 26-27, on Caribbean movements). Tabbarah (1981) noted the urban origin of migrants in the Arab region. A possible explanation to these distinct perceptions on the predominance of emigrants' rural or urban origins is found in the changing composition of the migrant population. Earlier studies tended to emphasize rural origin of migrants; recent ones highlight their urban background (Diaz-Briquets 1985 on Dominican emigrants; Alba-Hernàndez 1976, Portes 1979, and Verduzco Igartúa 1990 on Mexican emigrants).

Indications of a change in the socioeconomic origin[9] of Mexican migration suggest new insights on labor migration conceptualization. Individual migrant histories were perceived as paralleling the economies and societies they come from: large-scale urbanization, industrialization, and consolidation of strong and dynamic links of Mexico with the United States (Portes 1979).

This seminal analytical insight on Mexico has been extended to other regions, particularly Latin America, which has a higher urbanization than most of the less developed regions. A recent review on labor movements in the Americas challenges the view of the immigrant as a peasant and suggests that the origin of Latin American international migrants shifted from rural to urban in the last two decades (Pessar 1988, 3). This holds for migration within the region and for emigration from Latin America and the Caribbean to the United States (Balán 1985; Chaney 1985; Diaz-Briquets 1985).

Similar trends are present in Asian and African sending countries. The findings so far do not imply necessarily that the number of urban emigrants exceeds rural emigrants. What is clear is that the urbanization process appears in the composition of migration flows. When Peterson and Warren (1989) weighed "urban population growth rate" in exit countries on unauthorized migration to the United States,[10] this variable did not prove significant worldwide, but it was for Latin American and Caribbean countries.

In Latin America the urbanization process is often described as distorted in various ways—one being the high degree of primacy. In this context one relevant question needing full analysis is the mega-city role in international migration. Do these megalopolises act as substitute or complementary destinations for labor vis-à-vis international destinations? Richard Jones (1988) suggested there is greater emigration of Mexicans to the United States from areas far from domestic regional economic poles than from areas around them.

Changing Skills of Migrants

Many observers point out that current migrants are more skilled than earlier ones. Research in the Caribbean and South America points to a process of displacing skilled and semiskilled labor from stagnating sectors of national economies, particularly industry. During the 1970s, professionals and technical and other white-collar workers accounted for 70 percent of the employed emigrants who left Jamaica; significant losses of skilled workers also went unregistered. In certain categories, the unrecorded flows were nearly as large as or larger than the officially recorded outflows (Anderson 1988). Thomas-Hope (1985) argued that in most Caribbean countries, middle-class migration has greatly expanded. In Argentina, two flows can be distinguished: one that responded specifically to low salaries and scarce opportunities for progress of highly skilled personnel and another that emerged with the deterioration of the labor market for skilled and semiskilled workers (Balán 1985).

Regarding Asian flows toward the Middle East, Saith (1989) observed that the pattern of demand has shifted from several categories of unskilled and semiskilled workers—like those required in the domestic servant sector—to service, operations, and maintenance workers requiring higher skills. Among Asian labor flows toward the Middle East there already were important differences; some flows have been skilled (Koreans), some rather unskilled (Sri Lankans). Migration in Africa is also not limited to unskilled labor and often involves highly skilled workers (Adeokun 1991).

In general, between the migration of persons with limited skills (migrant workers) and highly qualified personnel (brain-drain), "the intermediate case, corresponding to skilled workers in construction and other trades, is becoming increasingly important" (Miró and Potter 1980, 136).

Many questions remain. Are the different compositions of migration flows and their changes over time related to the urbanization process in general or to the unique experiences of various countries? How can the phenomenon of rising skill levels of labor migrants be linked to continuing urban evolution or, conversely, to urban economic stagnation?

Given the current trend of rapidly increasing urbanization, what are the future prospects of migration? Must increasing urbanization translate to mounting pressures to emigrate? In many LDCs urbanization proceeds amid a growing extension of informal activities and the continuation of "traditional" networks of social relations. One would like to know from which sectors or social groups emigrants come—the formal sectors or the informal ones? In general, the relationship between development and migration is more complex than conventional wisdom considers it to be. Some aspects of development can encourage migration whereas others do not, and what happens in one country will not necessarily happen in another.

One basic conclusion to be drawn is that exit country populations are becoming increasingly urban, a fact certain to affect migrants' traits and their abilities to incorporate themselves into the destination society. Whether outside migration is related to the urban transition is a more difficult proposition. If so, economic and political conditions seem to play the lead role in migration. Here again changes in these conditions may have quite different implications in rural versus urban settings.

Effects of Labor Migration on Development

There is a dual rationale for the interest of labor-supplying countries in migration effects. First, an assumption exists that economic development would be accelerated by upgrading the labor force via temporary employment experience abroad and the application of acquired skills on migrants' return to the home country. Second, migration abroad is likely to facilitate greater integration of the sending country into the broader world economic system. Regarding the effects of emigration, attention focuses mainly on three aspects: demographic impacts, labor market effects, and the economic contribution of remittances.

Demographic Effects of Emigration and Return Migration

The most immediate effect of emigration is on the size and growth of the sending country population. First, the extent of this impact depends on the size of the emigrant population (net emigration) relative to the total. In small countries these effects could be considerable. For example, the demographic history of many small Caribbean island countries could not be understood without recognizing the major role of emigration. In Jamaica since the 1960s, net outflow seems to be more than half of the natural increase (Anderson 1988, 119). Even in large sending countries, where the relative size of the emigrant population is small but its distribution is uneven, important demographic impacts nevertheless may be felt in the population of the localities, communities, and regions where the bulk of the emigrants originate. This is especially important when population growth rates are declining (on Mexico, see García y Griego 1989).

Shifts in population growth rates caused by changes in international migration trends can be more sudden and abrupt than those caused by fertility and mortality changes. Tabbarah (1981, 183) notes the effects resulting from sudden shifts in the extent and direction of migratory flows in Jordan. Via a reversal in the direction of migratory flows, the population growth rate rose abruptly between 1975 and 1978 from 3 to 5 percent a year—a change in level from 1 percentage point lower to 1 percentage point higher than natural increase.

The demographic effects of labor migration exceed the simple equation of population growth; specific groups are generally more affected than others. Labor migration is age and gender selective; usually young adults and more men than women migrate. Thus, for temporary labor migration, the impact on population growth additionally depends on the balance of several factors—periods of spouses' separation, the extent of first marriage postponement, increase in migrants' income and wealth, and attitudinal changes caused by the migration experience. Migration effects on working-age groups have received a good deal of attention, usually tied to labor market considerations.

International migration trends may also have a significant impact on exit country population distribution.[11] If emigration originates mainly in rural areas, this feature may dampen domestic rural-to-urban trends; international and domestic destinations could act as substitutes. But if emigration originates mainly in urban centers, one could speculate that it creates a vacuum that in turn triggers an ever-intensified rural-to-urban migration.

Extending Todaro's model (1986) to international migration works under the assumption of a rural-urban-international continuum. It is usually assumed that flows of labor migration return (cyclically) to their origin. But the question of resettling returning migrants is becoming an issue of increasing concern for some countries. In this sense, one could formulate a preliminary distinction between the cyclical return of *migrant workers* (here I would include the permanent circulators and short-term contract workers in the Caribbean plus most Mexican migrants) and the resettlement processes of *return migration* (long-term Asian migrants, for instance). In the first case, the basic assumption would be that migrant workers do not change destination—rural to urban or vice versa—on their repeated returns. But this same assumption is less applicable regarding the destination of return migration.

Surveying the Asian situation, Saith (1989) concludes that although few urban migrants would return as rural residents, the reverse might not be true. Rural migrants might well try to make a transition into the economy's urban sector on their return. Usually, returnees go into small business and informal activities as the preferred avenues for reinsertion and resettlement. Urban areas are usually better endowed with infrastructure, so return migrants' businesses are mostly located there (Farooq-i-Azam 1987, 11 on an Asian Employment Programme study).

It is not surprising to find that return migrants choosing an urban location go to the largest cities, as did returnees from exile in South America. Most of the funding proposals for employment integration programs through job creation "were for micro businesses offering retail goods or services in the largest urban areas" (Marmora and Gurrieri 1988, 35). Unfortunately, the limited information on this matter does not let us discern the frequency and relevance of this shift in other settings and flows.

Labor Market Effects

Emigration is commonly perceived as a safety valve for mounting pressures in the labor market of supplying countries. Emigration purportedly siphons off a portion of the country's growing working-age population. Migration, even if temporary, is seen as alleviating the widespread un- and underemployment prevalent in these economies and somewhat relieving the downward pressures on wages attributed to an excessive supply of labor in the market. The Yemen Arab Republic is perhaps one of the most salient cases, with an estimated 30 percent of its labor force abroad, mostly in Saudi Arabia (Findlay 1987, 1). García y Griego (1989) estimated that the proportion of the Mexican-born work force resident in the United States was 6.1 percent in 1980; but this proportion is higher in the core regions where most emigrants come from.

This demographic effect of emigration on the working-age population does not translate directly into an unqualified relief function for the labor markets of sending countries. The short-term effects of reducing the working-age population are considerable, but the long-term and indirect effects have proved difficult to evaluate. Contrary to previous assumptions, unemployment is not the most common status of migrants.[12] In a survey in Mexico in the late 1970s (CENIET 1982), as many as 72 percent of the migrants who had worked in the United States reported some kind of economic activity in Mexico. Among workers away from their usual residence, only one in five did not work during the month preceding departure; the truly economically inactive unemployed accounted for a mere 3 percent of the whole sample.

This finding is repeated in research and reviews on Latin America (Diaz-Briquets 1983; Pastor 1985, 14) and probably everywhere. Stahl (1984), noting how Malaysia's manufacturing sector is plagued by skilled labor shortages at least in part because of labor emigration to Singapore, analyzes potential effects of emigration on the economies of sending countries. If emigrants can be replaced easily, there is no major disruption. If labor is highly varied in terms of skill, however, the situation is more disruptive. As skilled and unskilled workers constitute complementary factors, a shortage of one reduces the productivity and employment opportunities of the other, hence emigration could be harmful.

The issue of skills has resurfaced in the debate associated with the emerging trend of return migration in some Asian countries. The literature is generally quite critical on the issue of skill formation through migration. The theory of skill formation implicitly assumes that the set of jobs open to migrants in labor-importing countries requires a skill level exceeding the level the typical migrant initially had.[13] Linked with this theory is the assumption that emigrants will use the skills acquired abroad on their return.

Böhning (1984, 184), analyzing the Mediterranean experience, questions the "uncritical acceptance of the potential human capital gain from return migration" that assumes that reinsertion of migrants occurs in a formal, modern, and urban context. Saith (1989), surveying recent Asian experiences, concludes that "while some new skills might well have been imbedded overseas, these cannot be fully used upon return." There have been interpretations of the Mexican experience that migrants who exhibit higher skills and training (presumably gained abroad) tend to remain longer, or indefinitely abroad—in the United States (Alba 1985). Stahl (1984) observes that some migrant workers may prefer using their accumulated savings to set up businesses that bear little relation to skills they acquired abroad. Evidence from the Philippines indicates that most migrants do not plan to work in their previous occupation on return, let alone apply new skills acquired (Smart and Teodosio 1983).

Data on wage effects are contradictory. A review of intraregional migration concludes that under the labor conditions in most Latin American countries, effects of emigration on wages are likely to be imperceptible (Diaz-Briquets 1983). But Stahl (1984) cites a Thai study that found labor shortages due to emigration were sufficient to raise wages.

Insertion or reinsertion of return migration seems to take place largely in informal labor markets, "into so-called marginal tertiary activities" (Böhning 1984, 186). Many migrant returnees enter self-employment (see Dandler and Medeiros 1988 on Bolivian migrants, rural and urban, returning from Argentina), and the employment generation capacity of businesses set up by return migrants has been found to be limited. In Asia employment generation has been 1.4 additional jobs per unit (Farooq-i-Azam 1987, 12). But urban labor market effects cannot be dissociated from the issue of remittances. Research in five Colombian cities on the impact of the Venezuelan recession on migration found that adverse labor market conditions these cities experienced were more related to indirect effects caused by reduced economic activity because of diminished remittances than to the presence of returnees (Ungar Bleier 1988). In this case the assumption is that returnees just add to the labor supply.

Effects of Remittances

Remittances are a major contribution of labor migration to the supplying countries. Obviously their aggregate impact depends on their size and that of the country's economy, among other factors. In some cases, such as the Yemen Arab Republic, remittances are considered crucial (Findlay 1987). Ratios of remittances to exports have been calculated (Swamy 1981), as well as to exports and imports (Keely and Bao Nga Tran 1989), as an indication of their significance to the economies of exit countries. With the current deteriorating economic conditions and high debt service obligations of many labor-supplying nations, the contribution of remittances to some economies has been enhanced.

Issues related to remittances include their effects at the individual or household level and at the social or collective level. Remittances generally benefit the individual recipients, but this is not automatically extended to national or aggregate levels. Surveys from many countries (mostly in rural households with international migrant members) find that an important part of remittances (usually from two-thirds to four-fifths and more) is used for subsistence consumption (meeting day-to-day family needs) and the rest (usually a smaller part) goes into savings or some sort of investment.

In general, remittances from *circular,* short-term migrants and *prolonged,* long-term migrants seem different. In circular migration, the amounts of money sent home are regular and relatively small, and migrants do not return with large sums of money or major capital goods (for the Philippines, see Cariño 1987, 317-319; for the Caribbean, see Thomas-Hope 1985, 159; for traditional Mexican migration, see Alba 1985). In prolonged migration, in addition to the regular flow, physical and financial capital (savings) acquire particular relevance. In some Asian countries, policies encourage capital goods imports, and financial opportunities are offered to repatriate savings from abroad (on Pakistan, see Saith 1989).

The distinction between *regular remittances* and *financial capital* is important because the pattern differs depending on the migratory pattern. One would expect a greater proportion of expenditures on consumer durables with remittances received regularly than with capital brought in by migrants. Jamaican farm migrant households illustrate this point: A greater percentage of households save or invest with return capital than with remittances (Chaney 1988). Other dimensions are also relevant. Among Colombian migrants to Venezuela, only those with comparatively longer migratory experience and higher income levels can invest in economic activities, usually of an informal nature (Murillo and Silva 1984). In Mexico, the pattern of remittances expenditures is not fundamentally altered by the amounts remitted, family size, or number of wage earners in the household. The pattern changes slightly for migrants in a better socioeconomic position—the share going to immediate consumption diminishes slightly (see the review by Alba 1985).

In any case, there is a consensus that individual migrants, as well as their families, are undoubtedly better off. In Jamaica, Chaney (1988) asserts that households with international migrants hold a very clear advantage over the rest of the community. Cariño (1987) notes that Philippine migrant households have a higher socioeconomic status than non-migrant households—probably because of remittances and the need for a certain amount of household wealth to let a member make a long-distance move.

The literature reflects ambivalence about macroeconomic and social effects. On one hand, the benefits remittances represent are acknowledged, but on the other, some of their implications are regretted. A distinction must be made between local/community effects and regional/national ones. At the local level, although remittances often keep whole communities alive, this may have tragic results. The inflow of remittances substitutes an external source of income for an internal one, generating a sort of negative dependence. Agricultural production may cease; as Massey (1988, 400) put it: "In many cases the land simply is not farmed."

Similarly, the concentration of investment out of remittances in land and buildings is very often viewed as pushing up the price of agricultural land and the cost of construction materials and labor.[14] Some authors deplore such unfavorable social developments. Massey (1988, 399) cites Reichert's (1981) work on a rural Mexican town, observing that:

> Over two decades, migrant families representing 20 percent of those in the community gained control of 63 percent of the local land base, transforming the town *from an egalitarian community of universally poor, landless families* to one where economic power was concentrated in the hands of a landed migrant elite. (emphasis added)

For other authors, it seems not a minor achievement in itself that remittances help to maintain and reproduce established social and economic forms (Murillo and Silva 1984 on the border zones of Colombia with Venezuela; Chaney 1985 on the Caribbean islands).

The urban effects of remittances going directly into rural areas—where most studies have been done—usually are not specified. To grasp the importance of the urban effects, one needs to recall that the influx of remittances has important indirect effects in sectoral output and associated domestic employment by raising the aggregate capacity for labor absorption of the national economy. Habib (1985) found important output and employment increases for Bangladesh's economy using an input-output model. Although he did not break the model into urban and rural sectors, one can safely assume that in most exit countries important industry and services activities are urban. Similarly, although land and real estate purchases form a transfer, and as such the productive impact of these purchases is nil in direct terms, these transactions also would have "second-round effects," depending on the expenditure/saving/investment behavior of the person selling the property and receiving the transfer payment. In other words, real estate operations might not be quite so irrelevant (Saith 1989).

There is no general consensus on the impact of current international labor flows on sending countries. Most studies so far refer to farm households, peasant economies, and rural society. But what about urban settings? Do migrants return to urban destinations? How are remittances spent? What is the partition between consumption expenditures and savings in an urban context? Are remittances invested and consumed in similar or diverse shares compared to the rural economy and society? To which sectors and activities are the amounts invested bound? Do successful migrants (those with savings capacity and entrepreneurship) upon return move to the cities in search of investment opportunities? Recorded experiences, mostly from Asia, indicate that the labor market insertion of returnees to urban areas is mostly in informal activities. But perhaps the greatest "urban" impact of emigration is through its indirect effects on aggregate production and employment levels.

Final Comment

The urban factor in international labor migration has not received the attention it warrants. This bias is grounded in empirical evidence on prevailing circumstances and direct effects of labor migration in exit countries. Considering future trends of urbanization in these same countries, an argument could be made in favor of research that will pay more attention to the urban dimension of international labor emigration. Effects of labor emigration certainly weigh more heavily on urban areas than has been acknowledged to date.

Notes

1. Böhning (1983) denies that these features were intended or built into many West European guest worker programs (which excluded seasonal workers).
2. On the meaning of *residence* and *period of time*, see Zlotnik (1987).
3. The figures given here refer to migrant workers and exclude dependents living with them abroad, whose number is probably of similar magnitude. Further, the significant component of labor movements that goes undocumented or unrecorded frustrates estimates of the magnitude and changes of this phenomenon.
4. The invasion of Kuwait, the war that followed, and continuing unsettled conditions in and around the Persian Gulf impose too many questions on future trends of labor migration within that migratory system.

5. The underlying focus of research on emigration has so far been directed to rural conditions in exit countries. Urban aspects are more salient in studies dealing with conditions in destinations. It seems quite obvious that demand forces are predominantly urban in character and location. Further, large agglomerations and principal cities seem to be major destinations in developed as well as developing "receiving" nations.

6. A feature of labor movements associated with this factor is the character of legal or illegal, regular or irregular, documented or undocumented migration, and attendant implications.

7. Even within the perspective of world system interdependence, the reasons and explanations may greatly differ. See Böhning (1984, 133-143); Portes and Bach (1985, 1-28); Zolberg (1989).

8. Family and community networks play a significant role at both ends of the migration flow; for their relevance to immigration issues, see Portes and Böröcz (1989).

9. In the sense of having an urban rearing, background, or exposure rather that being urban by birth.

10. They consider there are insignificant behavioral differences regarding unauthorized and authorized migration. Their sample covers 69 countries.

11. For immigration, the literature shows generally the clustering of migrants in urban areas, with high concentration in large cities.

12. Similarly, it is found repeatedly that the migrants do not come from the poorest segments of the population.

13. A contrary view hypothesizes that some sort of "deskilling" could take place. Jobs open to migrant workers in the United States are generally unskilled, on the lower rungs of the job ladder, and offer little opportunity for advancement or skill acquisition (Piore 1979, Chapter 2).

14. Stahl (1984) observes that whether this increased demand manifests itself in inflationary pressures depends on supply elasticities of domestic industries and the foreign exchange regime the government adheres to.

12

Third World Urbanization, Migration, and Family Adaptation

S. PHILIP MORGAN

Rural-Urban Differences in the Family

This chapter focuses on the connections between one of the most powerful contemporary social processes, urbanization/industrialization,[1] and one of society's fundamental institutions, the family. How does urbanization affect the family and how, in turn, does the family influence urbanization and economic development? To address these questions, I review the research literature and published data to document differences across societies in urban and rural families and familial institutions. A preliminary question is: What is measured by these rural-urban contrasts?

A major weakness of much quantitative social science research is in interpreting observed differences as reflecting only secular change. This tendency arises because most theories focus on broad, long-term patterns of change, but most empirical evidence comes from cross sections or at best covers short, recent periods. Thus studies with short time horizons are used to test broad theories about societal change although idiosyncratic and temporary factors could account for many

---

AUTHOR'S NOTE: I acknowledge research support from the University of Pennsylvania and the Australian National University. I thank Samuel Preston, Allan Parnell, and Mari Clark for comments on an earlier draft.

changes. Differences also could be rooted in stable, fundamental differences in social structure and organization.

Social scientists have devoted no greater attention to any contrast than that between rural, traditional society and modern, urban, industrial society. The effects of urbanization on the family have been the focus of classic works such as Goode's *World Revolution and Family Patterns* (1963). Freedman (1961/1962, 56) summarizes the dominant theories succinctly:

> Industrial urbanization was associated with a much more complex division of labor in all spheres of life; together with the associated high rate of social and physical mobility, this inevitably led to a growth of secularism and rationalism, the declining influence of traditional forces such as religious faith, the shattering of the traditional family and other primary group associations, the growth of either anomic individualism or the attachment of the individual to a large impersonal, specialized organization. An essential element in this view of urban life was the idea that the family would lose its functions to other specialized institutions. Under these circumstances, children . . . would cease to be productive assets in a family-based economy and . . . would be impediments to active participation in the larger nonfamilial-based organization from which the rewards of an urban society would come. The dominant view was that as whole populations became involved in the urban market and society, family planning would become universal and the size of family planned would continue to decline. A logical extension of the basic premises of this model made it appear that the family would continue to decline among the major institutions, becoming eventually an unstable organization for satisfying the affectional needs of the couple. Under these circumstances completely new kinds of relationships for meeting sexual and affectional needs might develop.

This basic paradigm is powerful, has intuitive appeal, and fits the broad outlines of social change. Research of the past four decades has produced significant caveats, however, including a greater consideration of historical differences or variety in families, a greater appreciation for the persistence and strength of traditional family forms, and a greater recognition of how these first two factors allow for a range of family forms today. Families also confront short-term forces that may be very pressing and urgent. Families may or may not respond in a way that furthers their or society's long-term interests. Some rural-urban differences may be stable, rooted in fundamental differences in rural and urban life. In his classic essay on urbanism, Wirth (1938, 12) defines the *city* as a "relatively large, dense, and permanent settlement of socially heterogenous individuals." Based on sociological propositions about these demographic features, Wirth reasoned that social life in cities is basically different than it is in rural locales. For instance, large size implies that people cannot know one another; secondary (versus primary) contacts increase. Although Wirth acknowledges a positive side to the ledger, he clearly saw the balance as negative. Hawley (1972) tries for a more balanced view of the effects of urbanism, but he too sees fundamental structural differences by residence that influence individual behavior, including family behavior. He sees expanded opportunities as a fundamental feature of city life.

The concerns raised by rural-urban differences depend on one's interpretation of them. Freedman describes how the family becomes a weaker institution, giving up important functions, as a society develops (see Popenoe 1988). Some see this decline as the harbinger of a series of negative changes, eventually leading to societal collapse. Others see the decline as producing positive and negative consequences. The most common concerns focus on how the weakened family will continue to perform the crucial functions it retains: procreation, socialization, and economic support of children. Still others focus only on the perceived positive aspects of family decline (e.g., greater equality for women).

If short-run factors are relevant, then concern centers on how the family adjusts or responds. Do responses to short-run factors have long-term effects? For instance, temporary work migration may be an immediate answer to unemployment, but in the long run family stability may be undermined by the spousal separation and resulting altered roles. Studies with a Western focus on family response to crisis include historical (Hareven 1982; Modell 1978; Modell and Hareven 1973) and contemporary research (Elder 1974). Family and kin are often the first resource tapped in a crisis. Although anthropological work in Third World countries has addressed some of the same issues, it tends to focus on rural areas (see Bledsoe works cited in this review). Less work has been done in the rapidly growing urban areas of developed countries.

Family Adaptations to Urban Environments

Marriage, fertility, residence, and migration are family strategies[2] aimed at expanding family wealth, security, or social standing. We understand family change and these individual domains better when we recognize the possibility of a "multiphasic" response to similar exigencies (see Greenhalgh 1988). Family adaptations to urban environments can be expected in a set of distinct domains. Important historical and cultural factors in a given setting predispose families toward certain strategies, and institutions affect the type and magnitude of change (see Blumberg 1984; Folbre 1988). Societies tend to adapt to new circumstances with responses requiring the least institutional adjustment and personal sacrifice (see Davis and Blake 1956). Below I review research on a range of possible changes/adjustments families can make: changes in the timing of family formation, the number of children married couples have, living arrangements, intergenerational income flows, intrafamilial division of labor, and residence for better employment or other opportunities.

Marriage

In most 19th-century European populations marriage was earlier in urban than in rural areas. Late marriage in rural areas was tied to the land tenure system. Young couples were discouraged from marrying until they had acquired a means of support, usually access to farmland. Parents controlled farmland and did not release it to their progeny until their old age or death (Hajnal 1953). Working and living in cities removed this obstacle, allowing marriage at younger ages (Levine 1977).

In contrast, "in today's LDCs urban residence is all but universally associated with marriage delay" (Smith 1983, 499). In Table 12.1, I show percentages of women aged 15-19 who have never married by current residence[3] and men aged 20-24 who report never being married, using data for 24 developing countries that participated in the World Fertility Survey (WFS) program. Proportions never married at these ages vary sharply across countries: Virtually no rural Korean women have wed whereas only about one-fourth of Bangladesh rural women have never married. There is a consistent rural-urban differential. Column 3 in each panel shows rural-urban (absolute) differences. Negative numbers indicate that lower proportions have never wed in rural areas. Of these 24 countries, only two fail to show later marriage for urban women, although five do not show the expected pattern for men. The differential tends to be negative, but its magnitude varies considerably. Much of these observed differences could be attributed to secular factors of socioeconomic development.

Table 12.1 Percentage of Women Never Married at Ages 15-19 and Men Never Married at Ages 20-24 by Rural-Urban Residence and Country

	Women Never Married Ages 15-19			*Men Never Married Ages 20-24*		
	Rural	*Urban*	*Difference*	*Rural*	*Urban*	*Difference*
Africa						
Cameroon	47	59	−12	63	83	−20
Ivory Coast	48	55	−7	73	84	−11
Mauritania	60	63	−3	85	85	0
Morocco	71	88	−17	61	87	−26
Sudan	74	85	−11	81	92	−11
Asia and Pacific						
Jordan	71	84	−13	67	80	−13
Syria	76	78	−2	72	84	−12
Bangladesh	28	45	−17	60	71	−11
Nepal	40	72	−32	30	68	−38
Pakistan	58	72	−14	65	73	−8
Sri Lanka	93	94	−1	88	87	1
Fiji	87	89	−2	66	69	−3
Indonesia	58	80	−22	51	71	−20
Korea	99	99	0	95	97	−2
Malaysia	85	97	−12	77	90	−13
Philippines	92	96	−4	73	80	−7
Thailand	83	92	−9	61	79	−18
Americas						
Colombia	80	89	−9	75	75	0
Peru	81	88	−7	65	78	−13
Costa Rica	81	90	−9	68	72	−4
Dominican Republic	67	77	−10	69	68	1
Mexico	74	87	−13	54	64	−10
Panama	74	85	−11	68	70	−2
Haiti	82	77	5	79	60	19

SOURCE: Zoughlami and Allsopp 1985, Tables A11–A14.

Part of the later marriage in the urban areas can be traced to a greater number of alternative roles for young adults, particularly women's greater opportunities for schooling and work. Those who adopt these roles tend to marry later. Salaff (1976) relates how young, working,

Hong Kong women delayed marriage, with economic and social benefits for them and their parents.

Analyses have tried to test this hypothesis by comparing rural and urban women, holding constant factors such as years of schooling and work experience. If controls on opportunities, such as education, eliminated the rural-urban differences, then one would conclude that the major relevant feature of the urban-rural contrast was greater options. In fact, such controls do attenuate the rural-urban contrast, but they do not eliminate it (see Smith 1983, 499-501). This finding is consistent with the view that alternative roles are an important aspect of rural-urban contrasts.

Changes in the way marriages are arranged is a second contributing factor. Much evidence suggests that parents' roles in arranging marriages are declining (Thornton and Fricke 1987). Thornton, Chang, and Sun (1984) present strong evidence that this is so for Taiwan, showing that the erosion of parental prerogative is greatest in urban areas. Free choice of spouse does not guarantee later marriage. Even where parents play a large role in matchmaking, there are reasons to suspect that economic development or urbanization delays marriage. Based on their South India ethnosurveys, Caldwell, Reddy, and Caldwell (1983) maintain that arranging marriages becomes more complicated and takes longer when there are more gradations in status. An educated son needs an educated wife, for instance. These new considerations complicate and lengthen the search for an appropriate spouse. A parallel argument can be made in a freer marriage market. "Other determinants being the same, marriage should generally be later in dynamic, mobile, and diversified societies than in static, homogenous ones" (Becker 1973, S22). Longer searches are rational in more diverse populations because potential mates vary more.

As noted, existing institutions can influence the pace and form of change. Watkins (1984, 311) cites a Chinese ethnographer who said in the East "only the sick and morally depraved do not marry." But in the West, "spinsters were thick on the ground." Although there is a clear trend toward later marriage in Asian developing countries, signs of nonmarriage and childlessness are weak. Marriage comes later (with timing more like Western marriages), but it retains the traditional Eastern feature of being nearly universal.

A second important example of cultural continuity is the persistence of sub-Saharan African polygyny. Goode and others predicted that as societies become more developed, polygyny would decline. To the contrary, there is little evidence of a decline over recent decades

(Lesthaeghe, Kaufmann, and Meekers 1989). Even if African growth rates decline substantially, polygyny can be maintained as long as there are large age differences between husbands and wives and widows remarry (Goldman and Pebley 1989). Further, there are some advantages of polygyny in a modern context for men and women (Lesthaeghe, Kaufmann, and Meekers 1989).

Thus, despite higher age at marriage, distinctive cultural features are still embedded in contemporary rural and urban marriage patterns. Some patterns are adapted to the urban context. Several authors describe a new variant of polygyny in modern African cities (e.g., Dinan 1983).

Short-run factors also delay marriage. Caldwell, Reddy, and Caldwell (1983) report in their South India ethnosurveys that the major factor parents mention to explain why their daughters are not yet married is a shortage of appropriately aged spouses (also see Salaff 1976). Given husbands' substantially older age at marriage than wives and a rapidly growing population via mortality decline, the number of "marriageable" women will exceed the number of men. A shortage of spouses makes finding the right one more difficult regardless of who arranges the match. Marriage squeeze played a major role in recent increases in age at marriage, especially in Asia (Preston and Strong 1986). If fertility declines sharply, the imbalance can shift in favor of women. Marriage squeeze delayed marriage over the past few decades, but this is a transitional feature that might favor women's marriage prospects in the future.

Urban areas also have sex ratios far more unbalanced through selective migration than those produced by expanding age structures and depending on whether employment opportunities are greater for men or women. Opportunity varies across countries and over the course of development. Whether caused by rapid mortality decline or selective migration, sex ratios are best considered a transitional factor affecting urban environments.

A host of other short-run or idiosyncratic factors, such as a housing shortage, might delay marriage, but there are few data to examine this issue.

Marital Disruption. A priori, we expect greater marital disruption in urban areas where women's greater nonfamilial opportunities should be an important factor. Again using WFS data, I show the proportion aged 40-44 who are divorced or separated in rural and urban areas (Table 12.2). Results for women are very consistent. In 19 of 23 countries, a larger percentage of all women are divorced/separated in urban areas.[4]

Table 12.2 Percentage of Women and Men Divorced or Separated at Ages 40-44 by Rural-Urban Residence and Country

	Women			Men		
	Rural	*Urban*	*Difference*	*Rural*	*Urban*	*Difference*
Africa						
Cameroon	4	9	−5	5	4	1
Ivory Coast	5	7	−2	4	2	2
Mauritania	11	20	−9	2	3	−1
Morocco	3	5	−2	1	1	0
Sudan	6	6	0	2	2	0
Asia and Pacific						
Jordan	7	8	−1	1	1	0
Syria	1	2	−1	0	0	0
Bangladesh	1	2	−1	1	0	1
Nepal	2	5	−3	1	5	−4
Pakistan	2	3	−1	2	1	1
Sri Lanka	3	4	−1	1	1	0
Fiji	2	4	−2	2	0	2
Indonesia	7	9	−2	2	0	2
Korea	2	4	−2	2	1	1
Malaysia	6	4	2	2	0	2
Philippines	2	2	0	1	1	0
Thailand	3	10	−7	1	1	0
Americas						
Colombia	4	9	−5	1	2	−1
Peru	5	9	−4	2	2	0
Costa Rica	7	11	−4	4	5	−1
Dominican Republic	10	26	−16	9	9	0
Mexico	4	10	−6	2	2	0

SOURCE: Zoughlami and Allsopp 1985, Tables A11-A14.

The opposite is true for men: Rural areas have disproportionate numbers of separated/divorced men.

Although there is some evidence of higher divorce in urban areas, a greater proportion divorced (or widowed) in cities could come, not from divorce (or mortality) differentials, but by selective migration. If widows and divorcees have few opportunities for self-support in rural areas, migration to cities or towns for employment is one strategy for dealing with loss of a spouse. Historical and contemporary examples of such migration are given by Timaeus and Graham (1989) on female migration in Lesotho and Botswana, and Clark (1984) on women-headed households in Kenya.

Marital Fertility. Urban-rural fertility differentials are nearly ubiquitous. For most countries participating in the WFS program, the data show a clear gradient with fertility declining across rural, urban, and large city contexts. But as with marriage timing, remarkable variability exists.

Three broad sets of factors together produce the observed fertility differentials: secular, temporary, and stable. Using a multilevel analytic strategy and WFS data from 15 countries, Entwisle and Mason (1985) showed that residence effects on fertility vary by two features of the macro context: GNP per capita, measuring the secular socioeconomic trend, and the country's level of family planning effort, a temporary or episodic factor, not linked to secular change. Their important results follow:

	Family Planning Effort		
GNP per Capita	Low	Moderate	High
77	−.99	.08	.62
249	.07	.46	.65
421	1.14	.84	.68
593	2.21	1.21	.72

These values measure the amount by which rural fertility (e.g., lifetime children born to women aged 40-44) exceeds urban. At very low levels of development with little family planning effort, rural fertility is below urban fertility, consistent with evidence suggesting that fertility increases slightly before it begins to decline (Dyson and Murphy 1985). With economic development (increases in GNP per capita), the differential reverses and rural fertility exceeds urban. The differential is strongest where there is little family planning effort. Effects of city residence (compared to rural) are attenuated with a substantial family planning effort. Presumably, the weaker rural-urban contrast results because some advantages of city residence (greater accessibility to contraceptives) are neutralized by programmatic efforts.

Explanation for these differentials and their variability can be at two distinct levels. One identifies the proximate determinants (Bongaarts and Potter 1983). Socioeconomic variables, like urban residence, cannot influence fertility directly. Instead, residence operates by affecting the proportion of the population married, the length of postpartum infecundity, contraception, and abortion (p. 5). Using WFS data, one can document that urban areas, versus rural ones, generally have later marriage, more marital disruption, shorter postpartum infecundity (largely from shorter duration of breast-feeding), greater use of contraception, and although the evidence is weakest here, greater use of abortion. All of these changes reduce fertility, except reduced breast-feeding. The magnitude of the rural-urban differential is determined by the balance of these proximate determinants.

A second level of explanation looks for the more distant or fundamental reason why the proximate determinants, and indirectly, fertility vary. Less breast-feeding can be accounted for by available substitutes (formula) in urban areas or because breast-feeding is more compatible with rural women's work (Smith and Ferry 1984, Table 6).

Contraceptive knowledge and access is greater in urban areas (Jones 1984, 87). Differential motivation for additional children is also important in accounting for differential contraceptive use. Rural children are more productive than urban children, which may increase parents' demand for children (Becker 1981). Much work remains to be done in understanding fertility decisions. Work in the Third World has only begun to evaluate husbands and wives' separate desires and how they are reconciled (see Mason and Taj 1987).

As with increasing age at marriage, lower fertility does not necessarily destroy all important cross-cultural variability. Even low mortality/fertility regimes exhibit a range of fertility patterns. All women could become mothers, or fewer women could bear more children, producing an equally low mean number. Very large families are rare in developed countries, but other features of modern fertility patterns (e.g., childlessness and the sequencing of marriage and parenthood) continue to show substantial variability (Morgan, Rindfuss, and Parnell 1984).

The Household Division of Labor. The household division of labor varies by age and sex, and the degree of task segregation varies greatly across societies. Although direct evidence is scant, women's greater nonfamily roles in urban areas and with economic development[5] likely reduce differences between the roles and power of men and women (see Huber and Spitze 1983). Sociologists like Durkheim and economists like Becker (1981) argue that less task differentiation increases women's power relative to men and reduces conjugal solidarity. But political and institutional factors can affect the extent to which this occurs. As Folbre (1988, 258) argues:

> Elder men are able to control the ways in which women participate in wage labor, limiting their access to all but the lowest paying jobs. Employers, as

> well as men, benefit from low-cost female labor, and both occupational segregation and the sexual wage differential reproduce traditional patriarchal inequalities within modern capitalist systems.

Greater education, which comes with urbanization, has in most contexts eventually weakened these institutions.

Blumberg (1988; Blumberg and Clark 1989) argues that income under husbands' versus wives' control has consequences beyond marital solidarity because men and women spend income differently. Specifically, women hold back less for their own consumption than men do. Thus, increasing income under women's control has greater effects on dependents than does increasing men's. Blumberg's evidence is only suggestive and further evidence is needed.

Children's roles change drastically with economic development. With development and urbanization, children contribute less in work and their demands on household resources increase. Tienda (1979) shows that children aged 6-13 and 14-18 in Peru are far more likely to be working in rural than urban areas regardless of school attendance.

Although men, women, and children have normative roles, behavior frequently deviates from them when families confront crises. An immediate crisis may require wives to seek employment, and children may drop out of school and take a job to contribute to current family income. Increasing the number of family workers is one family response to crisis. Using data from urban areas in Colombia, Ecuador, and Peru, Musgrove (1980) shows that the poorest families had almost all possible members in the formal or informal labor force.

Living Arrangements: Family and Household Structure. Goode (1963) and many others argue that secular trends associated with economic development and urbanization would produce *conjugal families*—husband, wife, and children. A feature of this modern family is nuclear residence. Thus, as urbanization and economic development proceed, household complexity would increasingly be simple (nuclear) rather than complex (extended). With declining fertility and the desire for privacy, families and households would also be smaller.

As countries become more urbanized, household size declines largely through declining fertility (see UNDIESA 1980, 97-99). If one looks at rural-urban differences in contemporary developing countries, evidence is much weaker. In an analysis of 1960-1970 census data from 67 countries, the United Nations (UNDIESA 1980) reports smaller household size in the urban areas of all but 14 countries, but the differences tended to be very small. Data for a set of WFS countries show the same substantive result (Zoughlami and Allsopp 1985).

Evidence suggests a weak link between simpler family structure and urbanization. Primary evidence for an association comes from higher headship rates for adults in more urbanized countries and with urbanization in selected countries (e.g., Japan, Korea, and China; Martin 1990). But cross-sectional comparison of rural-urban differences shows little difference in household complexity. WFS data indicate that urban households are slightly *less* likely to be nuclear than rural households.[6] Clearly, this finding is counter to expectations generated from modernization theory. Detailed studies of living arrangements also show the persistence of extended living arrangements in urban areas (see Freedman et al. 1978; Martin and Tsuya 1989; Palmore, Klein, and Marsuki 1970).[7] Morgan and Hirosima (1983) show a high prevalence of extended residence among the most modern segments of contemporary Japanese society. Japanese subjects in this study saw important advantages in extended residence. Clearly extended residence can endure in an economically developed society, even in urban areas.

It would be a mistake to attribute all rural-urban differences to secular changes. Three sets of factors determine living arrangements: Economic constraints, availability of kin, and cultural predispositions. With a dominant rural-urban migrant flow, those remaining in rural areas likely have the largest set of kin with whom to live. But migration is often a chain movement based on kinship ties, so many recent urban immigrants might have relatives with whom they could live (see previous chapter).

Economic constraints and housing shortages can affect urban living arrangements. Housing costs can make nuclear residence prohibitively expensive. Sharing households as a response to economic crisis was common in the late 19th and early 20th centuries in the United States (Modell and Hareven 1973). Joining residences is probably a frequent response to periodic economic crisis in a great many contexts. Although Thai culture may be especially flexible in this regard, Chamratrithirong, Morgan, and Rindfuss (1988) show that couples chose from a range of residence options after marriage. Chinese Thais tended to live with the husband's parents and rural Thais with the wife's parents, but these "culturally preferred" patterns were not chosen if other arrangements offered substantial advantages. In urban areas, and especially in urban lower-class areas, recently married husbands and wives frequently lived apart because of the exigencies of work and shortage of housing.

Migration as a Family Strategy. The family plays a crucial role in migration by reducing both cost and uncertainty. My discussion relies heavily on work by Massey and his colleagues (1987).

Research has shown that moving where one has family, kin, or friends is a frequent strategy. Massey et al. describe how the migration flow from a rural community begins. First a few men migrate for seasonal, temporary work. If they are successful at finding work, they return to their community and tell others of their fortune. Others then repeat the migration. As these migrations become more regular and routine, women, wives, and children may accompany the men or may migrate alone. As migrants adjust to life at their destination, their stays may become longer. Eventually some become permanent residents in the new locale, and their returns to the village are now visits. As the community and family increase their ties in the destination, it becomes increasingly easy for subsequent individuals or families to move. Although details vary from place to place, the important point is viewing migration as a *process*. Family ties are central to it, creating a momentum to urban growth because the costs and uncertainty of migrating decline as the number of friends and relatives one has in cities of destination increases.

When the distance to urban areas from rural origins is less, then commuting to the city for work can become a long-term strategy. Hugo (1982) reviews evidence for Indonesia, describing an immense circular flow of individuals between Jakarta (and other major cities) and surrounding areas. Migrating for city employment has become so regular in parts of Indonesia that separate terms have developed for those who return nightly, those who return every few days, and those who migrate for several months. Circular migration, rather than becoming permanent as Massey et al. (1987) describe it, may be kept as a "risk aversion" strategy (see previous chapter). Given the uncertainty of city employment, rural ties (including residence) are considered important because support systems are more available there than in cities. Thus, if a "migrant maintains a stake in his village he does not cut himself off from what is often the only available support in times of dire need" (Hugo 1982, 71).

An interesting variant of the migration process in sub-Saharan Africa involves polygyny. Some men have a wife and family in the village and another in the city (Bledsoe forthcoming). The rural wife maintains the man's stake in the village and cares for his interests there. Another variant of these family strategies involves children. The institution of child fosterage has recently received substantial attention (see Isiugo-Abanihe 1985).

Page (1989) uses census data on relationship to household head to identify children who do not live in the same households as their mothers. In some areas of Cameroon, Ghana, Ivory Coast, and Lesotho, over 20 percent of children aged 0-14 do not live with their mothers. Older children are less likely to live with mothers in urban than in rural areas (see Page 1989, Figure 9.4), perhaps because disproportionate numbers of rural children are living with urban relatives because of the greater educational opportunities in the city.

Bledsoe and Isiugo-Abanihe (1989) claim that an educated family member with a job in government or the modern sector is a cherished asset. An urban contact in the modern bureaucracy is seen as vital for looking after family interests and as a sieve for sending future children into the modern sector. Again we see migration tying rural-urban families together and thus reducing the "social distance" rural migrants must traverse in moving to the city.

Classic Arguments Regarding the Family's Role in Facilitating/Hindering Urbanization

Above we considered family responses to the changed context urbanization presents. As Goode (1963) notes, the usual theory is that the conjugal family form emerges when urbanization and industrialization invade a culture. But he argues that the "family itself [must] be judged to have an independent effect on industrialization" (p. 22). He maintains that the conjugal family advances urbanization because it "fits" societies' changing needs. Goode claims that the nuclear household emerged at about the same time as industrialization and in response to the economic sector's demand for a socially and geographically mobile family. Urbanization involves job creation in new locations, and the conjugal family, with its small size and weaker ties to extended kin, is mobile geographically and socially and facilitates the movement of human resources to urban areas and into the economy's growing sectors.

The conjugal family encourages fewer children, allowing for greater investments in these progeny, especially in education. These more educated children fill crucial new niches in an expanding economy. The conjugal family is supported by an idealogy of individualism that allows for decisions based on universalistic criteria, presumed to be the most effective instrument for placing individuals in economic roles.

The question to be addressed now is: Have societies with other (nonnuclear) family systems been hindered by extensive kin ties and if not, how have these other systems facilitated or accommodated urbanization?

Do nonconjugal family forms hinder geographical mobility? Migration is a major contributor to city growth in developing countries. Much migration takes place as part of a family strategy or is eased by family contacts. Thus it is not clear that migration is easier for more nucleated families, as Goode maintains. Family ties can be important resources that make migration easier.

Do nonconjugal family forms hinder social mobility? Some argue that patriarchal, extended families retard social mobility because individual self-interest is moderated or thwarted by broader family interests. Levy (1968) argues that China's economic growth has been retarded by particularism, especially familism. But recent rapid economic development in Taiwan, South Korea, and other Asian countries challenges the view that strong, patriarchal, extended family systems thwart development. Also, extended residence increased in stages of industrialization in many Western countries (Ruggles 1987).

Do extended family ties maintain high fertility? The traditional view was yes. As Ronald Freedman (1979, 6) says about starting research in Taiwan: "17 years ago, it was expected that, if the wide-ranging development then underway continued, this would make for fundamental changes in the Chinese extended family structure, resulting in preferences for fewer children." High fertility was felt to slow development by reducing parental investment in each child (Blake 1981). High productivity and continued economic expansion depend on increasingly educated new cohorts. In Taiwan and several other Asian societies, the patriarchal extended family has remained largely intact although fertility has declined dramatically. Quoting Freedman (1979, p. 7) again:

> The Taiwanese apparently want to maintain intergenerational extended kinship ties in households equipped with the TV and all the latest electronic gadgets. They have rationally decided that this is feasible with fewer children in whom greater investment is made. . . . Large scale adoption of birth control and smaller families need not necessarily be preceded by the development of Western nuclear families.

Do extended families retard savings or capital formation? Future security may be enhanced more by strong family ties than by individual savings, increasing the likelihood that money will be spent maintaining strong ties rather than being saved. But the counterargument can be made as well: Capital for many ventures may be more easily accumulated by families than by individuals. Until other institutions for amassing capital have been developed, the role of the family and other informal associations could be crucial (McGee 1979).

In sum, the evidence fails to support the thesis that the nuclear family arose in conjunction with the economic development of the West and that it is uniquely suited to economically advanced society.

Familial Roles, Selective Migration, and Urbanization

Above, I argue that patriarchal and extended family systems can facilitate geographical mobility. The family contributes to city growth not only directly by expediting migration but also by the migrants' differential fertility and mortality. Also, migration is selective by family status and may produce populations more fitted to the demands of growing industrial sectors. If family and family roles played no part in migration, then rural and urban family statuses would be indistinguishable. Instead, among the laws of migration are such generalizations as "women predominate in short-term moves," thought to be accounted for by marriage. A second law is "single men predominate in long-term moves," presumably because young, unwed men have fewer family obligations and are freer to move. Younger workers may be more adaptable and trainable for new economic roles, and larger proportions of unmarried adults imply a lower urban dependency ratio. Are cities portrayed by such "favorable" demographic characteristics? Alternatively, selective migration may create problems. Widowed and divorced women might migrate to cities because of greater employment opportunities. Such families have high dependency ratios and high risk of being poor. I address two topics relevant to these issues: rural-urban variation in age structure and in sex ratio.

Age Structure Differences Between Rural and Urban Areas

Lower urban fertility makes urban age structure older. Migration, because it tends to select young adults, has the opposite effect. The balance of these two forces produces rural-urban age structure differences.[8] Overall, developing country urban areas have a younger age structure, more so where migration accounts for a greater proportion of urban growth (UNDIESA 1980).

Sex Ratios in Rural and Urban Areas

Rural-urban differences in sex ratios vary greatly by country. Rather extreme imbalances (beyond the 90:110 range) result from factors that induce one sex to migrate far more frequently. For countries with long time series data, evidence suggests cities were disproportionately male at the earliest date. The trend is toward more equal numbers of men and women (UNDIESA 1980).

Currently, in Africa, Oceania, and South Asia, cities tend to be more masculine than are rural areas (UNDIESA 1980, Figure 15). Although sparse, evidence suggests that these cities are becoming less male as well. If these areas follow the pattern of others, then the disproportionate number of men will decline as urbanization continues.

A new institution in Asia, dormitories for unmarried working women, may feminize cities earlier than would be expected otherwise. In Hong Kong and Taiwan, parents are often very reluctant to let their daughters migrate to cities. A high cultural value is placed on chastity until marriage and parents cannot exercise strong control if daughters do not live with them. Dormitories provided by employers are one solution (Thornton and Fricke 1987). Employers can recruit female workers for textile and electronics factories, and parents are more comfortable when their daughters live in supervised settings.

Latin America is unusual in that its cities are disproportionately female, no doubt because of large numbers of domestic servants in urban areas who live in the employer's household. Households in five Latin American capitals are more likely than rural ones to include individuals unrelated to the household head—these unrelated individuals are disproportionately unmarried women (DeVos 1987).[9]

Summary and Conclusion

Given the contrasts in traditional, agrarian society and modern, industrial society, family change is inevitable. In some cases, even the form and direction of family change seems determined. For instance, family size declines are inevitable with the increased cost of children in modern society. Teen marriage and childbearing are inconsistent with longer schooling. Such change indicates decline in the importance of the family, but even given these constraints, there is room for ample cross-cultural variability in family strength and structure. For instance, all men and women could marry and have children, or marital/parental roles could be more optional. Both strategies could satisfactorily meet requisite needs of replacement and socialization.

Consistent with this view are a number of constant differences between rural and urban areas that could be accounted for by secular, modernization factors. Marriage is later and fertility lower. Children's and women's roles change: Children are more likely to be in school, women more likely to be employed. But considerable variability remains, as in some countries the urban-rural contrast is dramatic and in others barely noticeable.

Urban contexts are highly variable and their effects should not be uniform. Work discussed here (especially by Entwisle and Mason 1985), indicates that the effects of urban contexts on fertility vary by level of economic development *and* the government's family planning effort. This finding is important because it challenges the notion of *an* effect of urbanization. For fertility at least (but also, we speculate, in a range of other domains) the effect of an urban context varies with economic development (a secular factor) and depends on a policy (an episodic factor).

There are strong reasons to expect that other macrolevel variables might also condition the experience of urban life, among them women's status and religion. Even in a fixed economic and policy context, groups with different historical and cultural backgrounds are likely to respond differently. Among many family strategies, such as adjusting the timing of marriage and fertility, some are more palatable to some groups than others. Some may be more useful in given situations or more advantageous to those in families with more power to influence family and individual decisions. Understanding the effects of urban context on the family requires a focus on historical and cultural contexts, idiosyncratic period factors, and the more commonly discussed secular changes.

Because my goal was to make international and cross-continent comparisons, this review has relied whenever possible on nationally representative data. The data allow for contrasts of rural and urban residents. Such a focus has severe shortcomings: It ignores differences *within* rural and urban populations; it overlooks the frequent migration between urban and rural areas that cannot be classified easily as rural or urban; and it suggests that migration decisions are individual ones.

To address the first issue, the substantial differences within rural and urban populations can be attributed to culture, socioeconomics, or policy. In developing countries, the growth of large squatter settlements may produce substantial variability by class or wealth in urban families. The strongest evidence comes from Bangkok, where Cherlin and Chamratrithirong (1988) show sharp differences in marriage patterns between established areas and squatter settlements where marriage occurred earlier, parents had a lesser role, and informal unions were more common. Much more such work needs to be done.

Although individual cities may contain substantial family diversity, urban contexts also vary across countries. Entwisle and Mason (1985) identify some dimensions along which cities vary that have implications for fertility. Other aspects of family behavior and dimensions of urban contexts must be examined. To understand better how urban contexts affect families, researchers must better measure the relevant features of this context, being sensitive to ways the effects of urban context vary by individual socioeconomic characteristics. Multilevel analysis (see Smith 1989) provides the appropriate analytic frame for these research questions.

The second shortcoming of the rural versus urban approach can also be overcome by collecting different, more detailed migration histories. Such histories have been used effectively in recent studies (Massey et al. 1987). These studies seem to show that frequent migrants more often adopt behavior and attitudes typical of their destination. Frequent migration could have other effects such as spousal separation lowering fertility.

The third problem arises from the way the data are collected and processed and the absence of a strong theoretical framework to guide the research. Tabulations are counts of individuals. It is easy to assume and to treat these counts as individual decisions to be correlated with other individual characteristics. But if migration is part of a family strategy, then treating it as an individual decision by the migrant misses much (see Caldwell 1985; Ruggles 1987; Ryder 1980). Empirical and theoretical attention must be devoted to the family context in which decisions are made. How are family decisions made, by whom, and for whose benefit?

Notes

1. I do not try to untangle effects of *urbanization* and *industrialization.* These changes are treated as inseparable aspects of economic development. Hereafter I refer only to urbanization.

2. A number of critiques attack the concept of family strategies (see Folbre 1988; Ruggles 1987, Chapter 2), particularly that families behave rationally to maximize the well-being of *all* members. Critics suggest that a chosen strategy may be in the best interest of none, or more likely, only some family members. For instance, men's and women's interests and goals may be very different. The critiques focus attention on how household decisions are made and, thus, on power relations within households. Although I use the terms *family adaptation* or *family strategy,* I make no assumption that decisions made are optimal for the family as a group or that they are equally beneficial to each family member.

3. World Fertility Surveys measured residence differently in various countries (see Singh 1980). Usually population size was used with other information in making the distinction, including whether a city had administrative functions for a region or the proportion involved in nonagricultural occupations. Some countries used a three-category variable (large city, other urban, rural) that I used in other tables.

4. Goldman's (1981) detailed analysis of WFS data for Colombia, Panama, and Peru shows sharp rural-urban differences in marriage and consensual union disruption. Evidence of higher divorce rates in urban (vs. rural) areas is weaker in Africa and Asia. Using WFS surveys for Ghana, Ivory Coast, and Nigeria, Brandon (1990) finds little difference in the risk of marital disruption for rural and urban women. But there are serious problems with age misstatement and reporting dates of marriage, separation, and divorce.

5. It is important to set the referent as traditional, agricultural settings, because there is much evidence that women's and men's roles and power were more equal in hunting and gathering societies than in agricultural societies (see Huber and Spitze 1983, Chapter 1).

6. Susan DeVos (1989) examines the likelihood that persons aged 15-29 would be reported as a child of the household head—a measure of the extendedness of households from the perspective of young adults. She uses WFS data for Colombia, Costa Rica, Dominican Republic, Mexico, Panama, and Peru. After controls for age, sex, and education, the "capital city, other urban and rural" differences were quite large: adolescents/young adults were much more likely to be living with parents in rural than capital city settings.

7. Further, studies show that more educated children are more likely to live with parents than those with less education (see DeVos 1989), which does not fit with the modernization view as education is frequently considered a key indicator of modernity.

8. Yi and Vaupel (1989) demonstrate the relative importance of these factors for China.

9. In Colombia and Costa Rica 20 percent of unmarried women aged 20-24 are unrelated to the head of the household in which they live (DeVos 1987, 516).

Bibliography

Abt Associates with Dames and Moore. n.d. *Informal housing in Egypt.* Cambridge, MA.

Abu-Lughod, Janet. 1975. Comments: The end of the age of innocence in migration theory. In *Migration and urbanization: Models and adaptive strategies,* ed. Brian M. Du Toit and Helen I. Safa, 201-206. Chicago: Aldine.

Abumere, S. I. 1982. Multinationals and industrialisation in a developing economy: The case of Nigeria. In *The geography of multinationals: Studies in the spatial development and economic consequences of multinational corporations,* ed. Michael J. Taylor and Nigel J. Thrift, 158-177. New York: St. Martin's.

Adeokun, Lawrence Adefemi. 1991. Population growth, migration and development in the African context. In *Consequences of rapid population growth in developing countries. Proceedings of the United Nations Expert Group Meeting, New York, 25-26 August.* ESA/P/WP.110. New York: Taylor & Francis.

Alba, Francisco. 1985. *Migrant workers, employment opportunities and remittances: The pattern of labor interchange between Mexico and the United States.* Occasional Paper Series, Center for Immigration Policy and Refugee Assistance. Washington, DC: Georgetown University.

Alba-Hernández, Francisco. 1976. Éxodo silencioso: La emigración de trabajadores mexicanos a Estados Unidos. *Foro Internacional* [Mexico City] 17:152-179.

Ambrose, William W., Paul R. Hennemeyer, and Jean-Paul Chapon. 1990. *Privatizing telecommunications systems: Business opportunities in developing countries.* IFC Discussion Paper no. 10. Washington, DC: World Bank.

Amin, A. T. M. Nurul. 1987. The role of the informal sector in economic development: Some evidence from Dhaka, Bangladesh. *International Labour Review* 126:611-623.

Anderson, Patricia Y. 1988. Manpower losses and employment adequacy among skilled workers in Jamaica, 1976-1985. In *When borders don't divide: Labor migration and refugee movements in the Americas,* ed. Patricia R. Pessar, 96-128. Staten Island, NY: Center for Migration Studies.

Arias, Patricia, and Bryan Roberts. 1985. The city in permanent transition: The consequences of a national system of industrial specialization. *Capital and labour in the urbanized world,* ed. John Walton, for the International Sociological Association, 149-175. Studies in International Sociology 31. London: Sage.

Arizpe, Lourdes. 1983. The rural exodus in Mexico and Mexican migration to the United States. *International Migration Review* 15:626-649.

Armstrong, Warwick, and Terence G. McGee. 1985. *Theatres of accumulation: Studies in Asian and Latin American urbanization.* London: Metheun.

Arn, Jack. 1987. *Dependency, underdevelopment and urbanization in the Philippines.* Working Paper no. 23. Hong Kong: Centre of Urban Studies and Urban Planning, University of Hong Kong.

Asian Development Bank. 1986. *Pakistan urban sector profile.* Draft for country review. Manila.

Balán, Jorge. 1985. *International migration in the Southern Cone.* Occasional Paper Series, Center for Immigration Policy and Refugee Assistance. Washington, DC: Georgetown University.

Bechhofer, Frank, and Brian Elliott. 1981. Petty property: The survival of a moral economy. In *The petite bourgeoisie: Comparative studies of the uneasy stratum,* ed. Frank Bechhofer and Brian Elliott, 182-200. New York: St. Martin's.

Becker, Gary S. 1973. A theory of marriage: Part II. *Journal of Political Economy* 82:S11-S26.

———. 1981. *A treatise on the family.* Cambridge, MA: Harvard University Press.

Beijing Review staff. 1988, 3-9 Oct. Begging becomes a popular profession. *Beijing Review* 31: 11-12.

Berg, Elliot, and Mary M. Shirley. 1987. Divestiture in developing countries. World Bank Discussion Papers no. 11. Washington, DC: World Bank.

Berry, Brian J. L. 1971. City size and economic development: Conceptual synthesis and policy problems, with special reference to South and Southeast Asia. In *Urbanization and national development,* ed. Leo Jakobsen and Ved Prakash, 111-155. South and Southeast Asia Urban Affairs Annuals, vol. 1. Beverly Hills, CA: Sage.

Berry, Brian J. L., and John D. Kasarda. 1977. *Contemporary urban ecology.* New York: Macmillan.

Berthelsen, John. 1989, 6 Sept. Thailand's bureaucrats, unions thwart government's programs of privatization. *Wall Street Journal/Europe,* p. 12.

Bhalla, A. S. 1973. The role of services in employment expansion. In *Third World employment: Problems and strategy. Selected readings,* ed. Richard Jolly et al., 287-301. Harmondsworth, UK: Penguin.

Bishop, Matthew R., and John A. Kay. 1989. Privatization in the United Kingdom: Lessons from experience. *World Development* 17:643-657.

Blake, Judith. 1981. Family size and the quality of children. *Demography* 18:421-442.

Bledsoe, Caroline H. Forthcoming. The politics of polygyny in Mende education and child fosterage transactions. In *Gender hierarchies,* ed. Barbara D. Miller. New York: Cambridge University Press.

Bledsoe, Caroline H., and Uche Charlie Isiugo-Abanihe. 1989. Strategies of child-fosterage among Mende grannies in Sierra Leone. In *Reproduction and social organization in Sub-Saharan Africa,* ed. Ron J. Lesthaeghe, 442-474. Berkeley and Los Angeles: University of California Press.

Blincow, Malcolm. 1986. Scavengers and recycling: A neglected domain of production. *Labour, capital and society* 19(1): 94-115.

Blitzer, Silvia, Jorge E. Hardoy, and David Satterthwaite. 1988. *The spatial distribution of aid for human settlements in the Third World: The role of multilateral agencies.* London: International Institute for Environment and Development.

Blumberg, Rae Lesser. 1984. A general theory of gender stratification. In *Sociological theory 1984,* ed. Randall Collins, 23-101. The Jossey-Bass Social and Behavioral Science Series. San Francisco: Jossey-Bass.

———. 1988. Income under female versus male control: Hypotheses from a theory of gender stratification and data from the Third World. *Journal of Family Issues* 9:51-84.

Blumberg, Rae Lesser, and Mari H. Clark, eds. 1989. *Making the case for the gender variable: Women and the wealth and well-being of nations.* Technical Reports in Gender and Development no. 1. Washington, DC: Office of Women in Development, Bureau for Program and Policy Coordination, U.S. Agency for International Development.

Böhning, Wolf-R. 1979. *Migration, the idea of compensation, and the international economic order.* ILO World Employment Programme Working Paper no. 2-26/WP.45. Geneva: International Labour Office.

———. 1983. Guestworker employment in selected European countries—Lessons for the United States? In *The border that joins: Mexican migrants and U.S. responsibility,* ed. Peter G. Brown and Henry Shue, 99-138. Maryland Studies in Public Philosophy. Totowa, NJ: Rowman & Littlefield.

———. 1984. *Studies in international labour migration.* New York: St. Martin's.

Bolderson, Claire. 1991. Privatisation adds to Indonesia's problems. *Financial Times* (17 July): 40.

Bongaarts, John, and Robert G. Potter. 1983. *Fertility, biology and behavior: An analysis of the proximate determinants.* New York: Academic Press.

Bornshier, Volker. 1980. Multinational corporations and economic growth: A cross-national test of the decapitalization thesis. *Journal of Development Economics* 7:191-210.

Boyd, Monica. 1989. Family and personal networks in international migration: Recent developments and new agendas. *International Migration Review* 23:638-670.

Brandon, Anastasia J. 1990. Marital dissolution, remarriage and childbearing in West Africa: A comparative study of Cote d'Ivoire, Ghana and Nigeria. Ph.D. diss., University of Pennsylvania.

Breman, Jan. 1976. A dualistic labour system? A critique of the "informal sector" concept. *Economic and Political Weekly* 11(48): 1870-1876; (49): 1905-1909; (50): 1939-1944.

Brennan, Ellen M., and Harry W. Richardson. 1989. Asian mega-city characteristics, problems, and policies. *International Regional Science Review* 12:117-129.

Bromley, Ray. 1978. Introduction—The urban informal sector: Why is it worth discussing? *World Development* 6:1033-1039.

———, ed. 1985. *Planning for small enterprises in Third World cities.* Oxford, UK: Pergamon.

Bromley, Ray, and Chris Gerry, eds. 1979. *Casual work and poverty in Third World cities.* Chichester, UK: John Wiley.

Business Asia staff. 1991. Privatization plan offers windows in Malaysia. *Business Asia* 23:88-89.

Caldwell, John C. 1985. Strengths and limitations of the survey approach for measuring and understanding fertility change: Alternative possibilities. In *Reproductive change in developing countries: Insights from the World Fertility Survey,* ed. John Cleland and John Hobcraft with Betty Dinesen, 45-63. London: Oxford University Press.

Caldwell, John C., P. H. Reddy, and Pat Caldwell. 1983. The causes of marriage change in South India. *Population Studies* 37:343-361.

Cameron, Gordon C., and Lowdon Wingo, eds. 1973. *Cities, regions and public policy.* Edinburgh: Oliver & Boyd for the University of Glasgow and Resources for the Future, Inc., Washington, DC.

Cariño, Benjamin V. 1987. The Philippines and Southeast Asia: Historical roots and contemporary linkages. In *Pacific bridges: The new immigration from Asia and the Pacific Islands,* ed. James T. Fawcett and Benjamin V. Cariño, 305-325. Staten Island, NY: Center for Migration Studies in association with the East-West Population Institute, East-West Center, Honolulu.

CASS. *See* Chinese Academy of Social Sciences.

Castells, Manuel. 1972. *La question urbaine.* Paris: F. Maspero.

———. 1977. *The urban question: A Marxist approach.* Trans. Alan Sheridan. Cambridge, MA: MIT Press.

Castells, Manuel, and Alejandro Portes. 1989. World underneath: The origins, dynamics, and effects of the informal economy. In *The informal economy: Studies in advanced and less developed countries,* ed. Alejandro Portes, Manuel Castells, and Lauren A. Benton, 11-37. Baltimore, MD: Johns Hopkins University Press.

CENIET. *See* Centro Nacional de Información y Estadísticas del Trabajo, Secretaria del Trabajo y Prevision Social.

Centre for Science and Environment, India. 1989. The environmental problems associated with India's major cities. *Environment and Urbanization* 1:7-15.

Centro Nacional de Información y Estadísticas del Trabajo, Secretaria del Trabajo y Prevision Social. 1982. *Los trabajadores Mexicanos en Estados Unidos. Resultados de la Encuesta Nacional de Emigración a la Frontera Norte del Pais y a los Estados Unidos. Mexico City.*

Chamratrithirong, Aphichat, Krittaya Archavanitkul, and Uraiwan Kanungsukkasem. 1979. *Recent migrants in Bangkok metropolis: A follow-up study of migrants' adjustment, assimilation, and integration.* Bangkok: Institute for Population and Social Research, Mahidol University.

Chamratrithirong, Aphichat, S. Philip Morgan, and Ronald R. Rindfuss. 1988. Living arrangements and family formation. *Social Forces* 66:926-950.

Chaney, Elsa M. 1985. *Migration from the Caribbean region: Determinants and effects of current movements.* Occasional Paper Series, Hemispheric Migration Project. Washington, DC: Center for Immigration Policy and Refugee Assistance, Georgetown University; and Geneva: Intergovernmental Committee for Migration.

———. 1988. *Migration, smallholder agriculture, and food consumption in Jamaica and Saint Lucia.* Washington, DC: Hemispheric Migration Project, Center for Immigration Policy and Refugee Assistance, Georgetown University.

Chapman, Murray. 1981. Policy implications of circulation: Some answers from the grassroots. *Population mobility and development: Southeast Asia and the Pacific,* ed. Gavin W. Jones and Hazel V. Richter, 71-87. Development Studies Centre Monograph no. 27. Canberra: Australian National University.

Chapman, Murray, and R. Mansell Prothero, eds. 1985. *Circulation in population mobility: Substance and concepts from the Malensian case.* London: Routledge & Kegan Paul.

Chase-Dunn, Christopher. 1975. The effects of international economic dependence on development and inequality: A cross-national study. *American Sociological Review* 40:720-738.

Cherlin, Andrew, and Aphichat Chamratrithirong. 1988. Variations in marriage patterns in central Thailand. *Demography* 25:337-353.

China State Statistical Bureau. 1988. *Tabulation of China 1% Population Sample Survey. National volume.* Beijing.

China Daily. 1988a, 10 Oct. Curbing migration of farmers to cities.

———. 1988b, 27 Oct. Vagrants increasing on city streets.

———. 1988c, 3 Nov. Capital's population reaches 10 million.

———. 1989a, 22 Feb. City staggers as countryside job hunters flood in.

———. 1989b, 4 Apr. Transients dodge family planning.

———. 1989c, 25 Nov. Prosperity for the people on the move.

———. 1990, 13 Mar. Best laid plans of Beijing go awry.

———. 1991a, 27 Feb. Transient family size targeted.

———. 1991b, 1 Mar. Guangdong sends outside job-seekers back home.

Chinese Academy of Social Sciences. 1988. *China. Migration of 74 cities & towns. Sampling survey data (1986). (Computer tabulation).* Beijing: Population Research Institute. (in Chinese)

Cho, Lee-Jay, and John G. Bauer. 1987. Population growth and urbanization: What does the future hold? In *Urbanization and urban policies in Pacific Asia,* ed. Roland J. Fuchs, Gavin W. Jones, and Ernesto del Mar Pernia with Sandra E. Ward, 15-37. Westview Special Studies on East Asia. Boulder, CO: Westview in association with the East-West Population Institute, East-West Center, Honolulu, and National Centre for Development Studies, Australian National University, Canberra.

Choi, Jin Ho. 1987. Republic of Korea country paper. In *Urban Policy Issues,* 477-527. Manila: Asian Development Bank.

Chowdhury, A., C. Kirkpatrick, and I. Islam. 1988. *Structural adjustment and human resource development in ASEAN.* New Delhi: International Labour Organisation.

Clark, Mari H. 1984. Women-headed households and poverty: Insights from Kenya. *Signs: Journal of Women in Culture and Society* 10:338-354.

Clark, Roger D. 1986. World-system position, economic development, and urban primacy: A cross-national study. *International Review of Modern Sociology* 16:1-17.

Cohen, Dennis J. 1985. The people who get in the way: Poverty and development in Jakarta. In *Planning for small enterprises in Third World cities,* ed. Ray Bromley, 219-232. Oxford, UK: Pergamon.

Cohen, R. B. 1981. The new international division of labor, multinational corporations and urban hierarchy. In *Urbanization and urban planning in capitalist society,* ed. Michael J. Dear and Allen J. Scott, 287-315. London: Methuen.

Conde, Julien, and Pap Diagne. 1986. *South-north international migrations: A case study, Malian, Mauritanian and Senegalese migrants from Senegal River Valley to France.* Paris: Development Centre of the Organisation for Economic Co-operation and Development.

Corbo, Vittorio, and Jaime de Melo. 1987. Lessons from the Southern Cone policy reforms. *World Bank Research Observer* 2:111-142.

Cowan, L. Gray. 1987. A global overview of privatization. In *Privatization and development,* ed. Steve H. Hanke, 7-15. San Francisco: International Center for Economic Growth, ICS Press.

Crenshaw, Edward M. 1991. Foreign investment as a dependent variable: Determinants of foreign investment and capital penetration in developing nations, 1967-1978. *Social Forces* 69:1169-1182.

Dandler, Jorge, and Carmen Medeiros. 1988. Temporary migration from Cochabamba, Bolivia, to Argentina: Patterns and impact in sending areas. In *When borders don't divide: Labor migration and refugee movements in the Americas,* ed. Patricia R. Pessar, 8-41. Staten Island, NY: Center for Migration Studies.

Davis, Kingsley, and Judith Blake. 1956. Social structure and fertility: An analytic framework. *Economic Development and Cultural Change* 4:211-235.

Deble, Isabelle, and Philippe Hugnon. 1982. *Vivre et survivre dans les villes Africaines.* Collection Tiers Monde. Paris: Presses Universitaires de France.

deGomez, M. Isabel, C. Ramirez, and A. Reyes. 1988. Employment and labour incomes in Colombia, 1976-85. In *Trends in employment and labour incomes: Case studies on developing countries,* ed. Wouter van Ginneken, 33-60. Geneva: International Labour Office.

DeHoog, Ruth Hoogland. 1984. *Contracting out for human services.* Economic, Political, and Organizational Perspectives. Albany: State University of New York Press.

DeJong, Gordon F., Brenda Davis Root, and Ricardo G. Abad. 1986. Family reunification and Philippine migration to the United States: The immigrants' perspective. *International Migration Review* 20:598-611.

dePardo, Monica Lanzetta, and Gabriel Murillo Castano with Alvaro Treana Soto. 1989. The articulation of formal and informal sectors in the economy of Bogotá, Colombia. In *The informal economy: Studies in advanced and less developed countries,* ed. Alejandro Portes, Manuel Castells, and Lauren A. Benton, 95-110. Baltimore, MD: Johns Hopkins University Press.

de Soto, Hernando. 1989. *The other path: The invisible revolution in the Third World.* Trans. June Abbot. New York: Harper & Row.

DeVos, Susan. 1987. Latin American households in comparative perspective. *Population Studies* 41:501-515.

———. 1989. Leaving the parental home: Patterns in six Latin American countries. *Journal of Marriage and the Family* 51:615-626.

Dewar, David, Alison Todes, and Vanessa Watson. 1986. *Regional development and settlement policy: Premises and prospects.* London: Allen & Unwin.

Diaz-Briquets, Sergio. 1983. *International migration within Latin America and the Caribbean: An overview.* CMS Occasional Papers and Documentation Series. Staten Island, NY: Center for Migration Studies.

———. 1985. Impact of alternative development strategies on migration: A comparative analysis. In *Migration and development in the Caribbean: The unexplored connection,* ed. Robert A. Pastor, 41-62. Boulder, CO: Westview.

Dietz, James L. 1986. Debt and development: The future of Latin America. *Journal of Economic Issues* 20:1029-1051.

Dinan, Carmel. 1983. Sugar daddies and gold-diggers: The white-collar single women in Accra. In *Female and male in West Africa,* ed. Christine Oppong, 344-366. London: Allen & Unwin.

Doebele, William A. 1987. Land policy. In *Shelter, settlement and development,* ed. Lloyd Rodwin, 110-132. Boston: Allen & Unwin.

Doeringer, Peter B. 1988. Market structure, jobs, and productivity: Observations from Jamaica. *World Development* 16:465-482.

Dogan, Mattei, and John D. Kasarda, eds. 1988a. *The metropolis era.* Vol. 1, *A world of giant cities.* Newbury Park, CA: Sage.

———, eds. 1988b. *The metropolis era.* Vol. 2, *Mega-cities.* Newbury Park, CA: Sage.

Donges, Juergen B. 1987. International migration and the international division of labor. In *Population in an interacting world,* ed. William Alonso, 129-148. Cambridge, MA: Harvard University Press.

Dowall, David E. 1989. *Bangkok: A profile of an efficiently performing housing market.* Institute of Urban and Regional Development Working Paper no. 493. Berkeley: University of California at Berkeley.

Drakakis-Smith, David W. 1987. *The Third World city.* London: Methuen.

Dunning, J. H. 1988. The eclectic paradigm of international production: A restatement and some possible extensions. *Journal of International Business Studies* 19:1-31.

Dupont, Veronique. 1989. Contribution of anthropological approach to migration study: The case of temporary industrial labour migration in India. Paper presented at the International Population Conference, Session F.25, September, New Delhi.

Dyson, Tim P., and Michael J. Murphy. 1985. The onset of fertility transition. *Population and Development Review* 11:399-440.

Eckholm, Eric P. 1977. *The picture of health: Environmental sources of disease.* New York: Norton.

Edgren, Gus, and M. Muqtada. 1989. *Strategies for growth and employment in Asia: Learning from within.* New Delhi: International Labour Organisation Asian Employment Programme.

Elder, Glen H., Jr. 1974. *Children of the Great Depression: Social change in life experience.* Chicago: University of Chicago Press.

El Sammani, Mohamed O., Mohamed El Hadi Abu Sin, M. Talha, B. M. El Hassen, and Ian Haywood. 1989. Management problems of Greater Khartoum. In *African cities in crisis: Managing rapid urban growth,* ed. Richard E. Stren and Rodney R. White, 247-275. Boulder, CO: Westview.

Entwisle, Barbara, and William M. Mason. 1985. Multilevel effects of socioeconomic development and family planning programs on children ever born. *American Journal of Sociology* 91:616-649.

ESCAP. *See* United Nations Economic and Social Commission for Asia and the Pacific.

Espenshade, Thomas J. 1988. Projected imbalances between labor supply and labor demand in the Caribbean Basin: Implications for future migration to the United States. Washington, DC: Urban Institute.

Faissol, Speridao. 1986. Brazil's urban system in 1980: Basic dimensions and spatial structure in relation to social and economic development. *The metropolis in transition,* ed. Ervin Y. Galantay, New York: Paragon.

Farbman, Michael, ed. 1981. *The PISCES studies: Assisting the smallest economic activities of the urban poor.* Washington, DC: Office of Urban Development, U.S. Agency for International Development.

Farooq-i-Azam. 1987. *Re-integration of return migrants in Asia: A review and proposals.* New Delhi: ILO-ARTEP.

Ferris, James, and Elizabeth Graddy. 1986. Contracting out: For what? With whom? *Public Administration Review* 46: 332-344.

Fields, Gary S. 1979. Place-to-place migration: Some new evidence. *Review of Economics and Statistics* 61:21-32.

———. 1988. Employment and economic growth in Costa Rica. *World Development* 16:1493-1509.

———. 1989. The impact of government policies on urban employment in small economies. In *Fighting urban unemployment in developing countries,* ed. Bernard Solomé, 105-121. Paris: Organisation for Economic Cooperation and Development.

Findlay, Allan M. 1987. *The role of international labour migration in the transformation of an economy: The case of the Yemen Arab Republic.* International Migration for Employment Working Paper no. 35. Geneva: International Labour Office.

Findley, Sally Evans. 1981. Rural development programmes: Planned versus actual migration outcomes. In *Population distribution policies in development planning; Papers of the United Nations/UNFPA Workshop on Population Distribution Policies in Development Planning, Bangkok, 4-13 September 1979,* 144-166. Population Studies no. 75. ST/ESA/SER.A/75. New York: Department of International Economic and Social Affairs, United Nations.

———. 1982. *Migration survey methodologies: A review of design issues.* IUSSP Papers no. 20. Liège, Belgium: International Union for the Scientific Study of Population.

———. 1991. Changing pace and scale of migration and urbanization in the developing world. Paper presented at the Symposium on Humankind and Global Change, American Association for the Advancement of Science, Washington, DC.

Finnegan, Gregory, A. 1980. Employment opportunity and migration among the Mossi of Upper Volta. *Research in Economic Anthropology* 3:291-322.

Folbre, Nancy. 1988. The black four of hearts: Toward a new paradigm of household economics. In *A home divided: Women and income in the Third World,* ed. Daisy Dwyer and Judith Bruce, 248-262. Stanford, CA: Stanford University Press.

Forbes, Dean, and Nigel J. Thrift. 1987. International impacts on the urbanization process in the Asian region: A review. In *Urbanization and urban policies in Pacific Asia,* ed. Roland J. Fuchs, Gavin W. Jones, and Ernesto del Mar Pernia with Sandra E. Ward, 67-89. Westview Special Studies on East Asia. Boulder, CO: Westview in association with the East-West Population Institute, East-West Center, Honolulu, and National Centre for Development Studies, Australian National University, Canberra.

Frank, Andre Gunder. 1966. The development of underdevelopment. *Monthly Review* 18(Sept.): 17-31.

Freedman, Ronald C. 1961/1962. The sociology of human fertility: A trend report and bibliography. *Current Sociology* 10/11:35-68.

———. 1979. Theories of fertility decline: A reappraisal. *Social Forces* 58:1-17.

Freedman, Ronald C., Baron L. Moots, Te-Hsiung Sun, and Mary Beth Weinberger. 1978. Household composition and extended kinship in Taiwan. *Population Studies* 32:65-80.

Frey, R. Scott, Thomas Dietz, and Jane Marte. 1986. The effect of economic dependence on urban primacy. A cross-national panel analysis. *Urban Affairs Quarterly* 21:359-368.

Friedmann, John. 1968. The strategy of deliberate urbanization. *Journal of the American Institute of Planners* 34:364-373.

———. 1986. The world city hypothesis. *Development and Change* 17:69-83.

Friedmann, John, and Goetz Wolf. 1982. World city formation: An agenda for research and action. *International Journal of Urban and Regional Research* 6:309-344.

Fuchs, Roland J., and George J. Demko. 1979. Population distribution policies in developed socialist and Western nations. *Population and Development Review* 5:439-467.

Fuchs, Roland J., Gavin W. Jones, and Ernesto del Mar Pernia, eds., with Sandra E. Ward. 1987. *Urbanization and urban policies in Pacific Asia.* Westview Special Studies on East Asia. Boulder, CO: Westview in association with the East-West Population Institute, East-West Center, Honolulu, and National Centre for Development Studies, Australian National University, Canberra.

Furedy, Christine. 1989. Social considerations in solid waste management in Asian cities. *Regional Development Dialogue* 10(3): 13-43.

Gakenheimer, Ralph, and Carlos Henrique Jorge Brando. 1987. Infrastructure standards. In *Shelter, settlement and development,* ed. Lloyd Rodwin, 133-150. Boston: Allen & Unwin.

Galantay, Ervin Y., ed. 1986. *The metropolis in transition.* New York: Paragon.

García y Griego, Manuel. 1983. The importance of Mexican contract laborers to the United States 1984-1964: Antecedents, operation, and legacy. In *The border that joins: Mexican migrants and U.S. responsibility,* ed. Peter G. Brown and Henry Shue, 49-98. Maryland Studies in Public Philosophy. Totowa, NJ: Rowman & Littlefield.

———. 1989. Mexico's "labor force safety valve" after IRCA: A first approximation. Paper read at the Conference on the International Effects of the Immigration Reform and Control Act of 1986, 3-5 May, Guadalajara, Mexico.

GEMS. *See* Global Environmental Monitoring System.

Gilbert, Alan G. 1974. Industrial location theory: Its relevance to an industrialising nation. In *Spatial aspects of development,* ed. B. S. Hoyle, 271-290. London: John Wiley.

———. 1976. The arguments for very large cities reconsidered. *Urban Studies* 13:27-34.

———. 1990. Land and shelter in mega-cities: Some critical issues. Paper presented at the symposium on the Mega-City and the Future: Population Growth and Policy Responses, Tokyo.

Gilbert, Alan G., and Josef Gugler. 1982. *Cities, poverty, and development: Urbanisation in the Third World.* Oxford, UK: Oxford University Press.

Gladwell, J. S. 1989. Urbanization, hydrology, water management and the international hydrological programme. Paper read at the Conference on Integrated Water Management and Conservation in Urban Areas, Nagoya, Japan.

Global Environmental Monitoring System. 1988. *Global pollution and health. Results of health-related environmental monitoring. Assessment of urban air quality.* Geneva: United Nations Environment Programme and World Health Organization.

Goldman, Noreen. 1981. Dissolution of first unions in Colombia, Panama, and Peru. *Demography* 18:659-679.

Goldman, Noreen, and Anne Pebley. 1989. The demography of polygyny in Sub-Saharan Africa. In *Reproduction and social organization in Sub-Saharan Africa,* ed. Ron Lesthaeghe, 212-237. Berkeley: University of California Press.

Goldstein, Alice. 1988. Temporary migration in Southeast Asia and China: New forms of traditional behavior. *American Asian Review* 6(2): 1-30.

Goldstein, Alice, and Sidney Goldstein. 1990. China's labor force: The role of gender and residence. *Journal of Women and Gender Studies* [Taiwan] 1:87-118.

Goldstein, Alice, Sidney Goldstein, and Shenyang Guo. 1991. Temporary migrants in Shanghai households. *Demography* 28:275-291.

Goldstein, Alice, and Shengzu Gu. 1991. Determinants of mobility strategy: The experience of Hubei Province. Paper read at the annual meeting of the Population Association of America, 21-23 March, Washington, DC.

Goldstein, Alice, and Shenyang Guo. 1991. Temporary migration in Shanghai and Beijing. Paper read at the annual meeting of the Population Association of America, 21-23 March, Washington, DC.

Goldstein, Sidney. 1978. *Circulation in the context of total mobility in Southeast Asia.* Papers of the East-West Population Institute no. 53. Honolulu: East-West Center.

———. 1990. Urbanization in China 1982-87: Effects of migration and reclassification. *Population and Development Review* 16:673-701.

Goldstein, Sidney, and Alice Goldstein. 1979. Types of migration in Thailand in relation to urban-rural residence. In *Economic and demographic change: Issues for the 1980's; Proceedings of the conference, Helsinki, 1978.* Vol. 2, 351-375. Liège, Belgium: International Union for the Scientific Study of Population.

———. 1981. *Surveys of migration in developing countries: A methodological review.* Papers of the East-West Population Institute no. 71. Honolulu: East-West Center.

———. 1986. *Migration in Thailand: A twenty-five year review.* Papers of the East-West Population Institute no. 100. Honolulu: East-West Center.

———. 1987. Migration in China: Data, policies, and patterns. Brown University, Providence, RI. Typescript.

Golini, Antonio, and Bonifazi, Corrado. 1987. Demographic trends in international migration. In *The future of migration,* 110-136. Paris: Organisation for Economic Cooperation and Development.

Golladay, Frederick, and Timothy King. 1979. Social development. In *Korea: Policy issues for long-term development: The report of a mission sent to the Republic of Korea by the World Bank,* ed. Parvez Hasan and D. C. Rao, 135-202. Baltimore, MD: Johns Hopkins University Press.

Goode, William J. 1963. *World revolution and family patterns.* [New York]: Free Press of Glencoe.

Gordon, D. L. 1978. *Employment and development of small enterprises.* Sector Policy Paper. Washington, DC: World Bank.

Gould, William Taylor Sparkie. 1976. *Longitudinal studies of population mobility in tropical Africa.* African Population Mobility Project Working Paper no. 29. Liverpool, UK: University of Liverpool.

Greenberg, Michael R. 1986. Disease competition as a factor in ecological studies of mortality: The case of urban centers. *Social Science and Medicine* 23:929-934.

Greenhalgh, Susan. 1988. Fertility as mobility: Sinic transitions. *Population and Development Review* 14:629-674.

Griffin, K., and A. K. Ghose. 1979. Growth and impoverishment in the rural areas of Asia. *World Development* 7:361-383.

Guest, Philip. 1989. *Labour allocation and rural development: Migration in four Javanese villages.* Boulder, CO: Westview.

Gugler, Josef, and Flannagan, William G. 1978. Urban-rural ties in West Africa: Extent, interpretation, prospects and implications. *Migration and the transformation of modern African society,* ed. William M. J. Van Binsbergen and Henk A. Meilink. Leiden, Netherlands: Afrika Studiecentrum.

Gwynne, Robert N. 1985. *Industrialization and urbanization in Latin America.* London: Croom Helm.

Habib, Ahsanul. 1985. Economic consequences of international migration for sending countries: Review of evidence from Bangladesh. Ph.D. diss., University of Newcastle, Australia.

Hahn, Yeong-Joo. 1989. Manufacturing changes and industrial location policies in Korea. Ph.D. diss., Syracuse University.

Hainsworth, Geoffrey B. 1990. Indonesia: On the road to privatization? *Current History* 89:121-124, 134.

al-Haj, Dato' Seri Radin Soenarno, and Zainal Aznam Yusof. 1985. The experience of Malaysia. In *Privatisation policies, methods and procedures,* 5-26. Manila: Asian Development Bank.

Hajnal, John. 1953. Age at marriage and proportions marrying. *Population Studies* 7:111-136.

Hamer, Andrew Marshall. 1983. *Decentralized urban development and industrial location behavior in São Paulo, Brazil: A synthesis of research issues and conclusions.* Washington, DC: World Bank.

Hanke, Steve H., ed. 1987. *Privatization and development.* San Francisco: International Center for Economic Growth, ICS Press.

Hansen, Niles. 1981. A review and evaluation of attempts to direct migrants to smaller and intermediate-sized cities. In *Population distribution policies in development planning; Papers of the United Nations/UNFPA workshop on Population Distribution Policies in Development Planning, Bangkok, 4-13 September 1979,* 113-127. Population Studies no. 75. ST/ESA/SER.A/75. New York: Department of International Economic and Social Affairs, United Nations.

Hardoy, Jorge E., and David Satterthwaite. 1981. *Shelter, need and response: Housing, land, and settlement policies in seventeen Third World nations.* Chichester, UK: John Wiley.

———. 1986. Urban change in the Third World. Are recent trends a useful pointer to the urban future? *Habitat International* 10(3): 33-52.

Hareven, Tamara K. 1982. *Family time and industrial time: The relationship between the family and work in a New England industrial town.* Cambridge, UK: Cambridge University Press.

Harper, Malcolm, and Tan Thiam Soon. 1979. *Small enterprises in developing countries: Case studies and conclusions.* London: Intermediate Technology.

Harpham, Trudy, Tim Lusty, and Patrick Vaughan. 1988. *In the shadow of the city: Community health and the urban poor.* Oxford, UK: Oxford University Press.

Harris, Nigel. 1978. *Economic development, cities and planning: The case of Bombay.* Bombay: Oxford University Press.

Hart, J. Keith. 1973. Informal income opportunities and urban employment in Ghana. *Journal of Modern African Studies* 11:61-89.

Harvey, David. 1975. The political economy of urbanization in advanced capitalist societies: The case of the United States. In *The social economy of cities,* ed. Gary Gappert and Harold M. Rose, 119-163. Urban Affairs Annual Reviews, vol. 9. Beverly Hills, CA: Sage.

Hatry, Harry P. 1983. *A review of private approaches for delivery of public services.* Washington, DC: Urban Institute Press.

Hawley, Amos H. 1972. Population density and the city. *Demography* 9:521-529.

Henderson, J. Vernon. 1986. Efficiency of resource usage and city size. *Journal of Urban Economics* 19:47-70.

———. 1988. *Urban development: Theory, fact, and illusion.* New York: Oxford University Press.

Hiebert, Murray. 1991. Vietnam: Private clinics introduced for flagging health system. The cost of care. *Far Eastern Economic Review* 151(Jan. 10): 16-17.

Hirst, Paul Q. 1975. Marx and Engels on law, crime and morality. In *Critical criminology,* ed. Ian Taylor, Paul Walton, and Jock Young, 303-332. London: Routledge & Kegan Paul.

Hong, S. W. 1987. The domains of regional policy. In *Urban and regional policies in Korea and international experiences,* ed. Harry W. Richardson and Myong-Chan Hwang, 59-104. Seoul: Kon-Kuk University Press in association with Korea Research Institute for Human Settlements.

Hope, Kempe Ronald. 1984. Unemployment, labour force participation, and urbanization in the Caribbean. *Review of Regional Studies* 14(3): 9-16.

Hoselitz, Berthold Frank. 1955. Generative and parasitic cities. *Economic Development and Culture Change* 3:278-294.

———. 1960. *Sociological aspects of economic growth.* Glencoe, IL: Free Press.

Hosier, Richard H. 1987. The informal sector in Kenya: Spatial variation and development alternatives. *Journal of Developing Areas* 21:383-402.

Huber, Joan, and Glenna Spitze. 1983. *Sex stratification: Children, housework, and jobs.* New York: Academic Press.

Hugo, Graeme J. 1978. *Population mobility in West Java.* Yogyakarta, Indonesia: Gadjah Mada University Press.

———. 1981. Road transport, population mobility and development in Indonesia. In *Population mobility and development: Southeast Asia and the Pacific,* ed. Gavin W. Jones and Hazel V. Richter, 355-381. Development Studies Centre Monograph no. 27. Canberra: Australian National University.

———. 1982. Circular migration in Indonesia. *Population and Development Review* 8:59-83.

———. 1987. Demographic and welfare implications of urbanization: Direct and indirect effects on sending and receiving areas. In *Urbanization and urban policies in Pacific Asia,* ed. Roland J. Fuchs, Gavin W. Jones, and Ernesto del Mar Pernia with Sandra E. Ward, 136-165. Westview Special Studies on East Asia. Boulder, CO: Westview in association with the East-West Population Institute, East-West Center, Honolulu, and National Centre for Development Studies, Australian National University, Canberra.

Hymer, Stephen 1975. The multinational corporation and the law of uneven development. In *International firms and modern imperialism: Selected readings,* ed. Hugo K. Radice, 37-62. Penguin Modern Economics Readings. Harmondsworth, UK: Penguin.

ILO. *See* International Labour Office.

International Labour Office. 1970. *Towards full employment: A programme for Colombia, prepared by an inter-agency team organised by the International Labour Office.* Geneva.

———. 1972. *Employment, incomes and equality: A strategy for increasing productive employment in Kenya.* Geneva.

———. 1986. *Economically active population: Estimates and projections, 1950-2025.* 3rd ed. Geneva.

———. 1987. *World labour report.* Vol. 3, *Incomes from work: Between equity and efficiency.* Oxford, UK: Oxford University Press.

International Labour Organisation. 1986. *Creation of productive employment: A task that cannot be postponed.* Santiago, Chile: ILO-Programa Regional del Empleo para America Latina y el Caribe.

———. 1987. *Structural adjustment: By whom, for whom: Employment and income aspects of industrial restructuring in Asia.* New Delhi: International Labour Organisation Asian Employment Program.

International Monetary Fund. 1985. *Foreign private investment in developing countries, a study by the research department.* Occasional Papers no. 33. Washington, DC.

Isiugo-Abanihe, Uche C. 1985. Child fosterage in West Africa. *Population and Development Review* 11:53-73.

Jansen, J. C., and Jean H. P. Paelinck. 1981. The urbanisation phenomenon in the process of development: Some statistical evidence. In *The dynamics of urban development,* ed. L. H. Klaasen, W. T. M. Molle, and Jean H. P. Paelinck, 31-46. New York: St. Martin's.

Japan External Trade Organization. 1986. *1986 JETRO white paper on world and Japanese overseas direct investment. Summary. Japan's overseas direct investment entering a vigorous new phase of internationalization.* Tokyo.

———. 1987. *1987 JETRO white paper on world and Japanese overseas direct investment. Summary. Japan's overseas investment entering a new phase with the yen's appreciation.* Tokyo.

———. 1989. *1989 JETRO white paper on world direct investments: New phase in foreign direct investments and strategic alliances.* Tokyo.

JETRO. *See* Japan External Trade Organization.

Jimenez, R. D., and A. Velasquez 1989. Metropolitan Manila: A framework for its sustained development. *Environment and Urbanization* 1:51-58.

Johnson, D. Gale. 1988. Economic reform in the People's Republic of China. *Economic Development and Cultural Change* 36:S225-S245.

Johnson, Edgar Augustus Jerome. 1970. *The organization of space in developing countries.* Cambridge, MA: Harvard University Press.

Jones, Elise F. 1984. The availability of contraceptive services. *World Fertility Survey Comparative Studies,* no. 37.

Jones, Gavin W. 1989. Structural economic change and its relationship to urbanization and population distribution policies. Paper read at the International Workshop on Urbanization and Population Distribution Policies in Asia, Honolulu.

Jones, Richard C. 1988. Micro source regions of Mexican undocumented migration. *National Geographic Research* 1:11-22.

Jules-Rosette, Bennetta. 1985. The women potters of Lusaka: Urban migration and socioeconomic adjustment. *African migration and national development,* ed. Beverly Lindsay, 82-112. University Park, PA: Pennsylvania State University Press.

Kasarda, John D., and Edward Crenshaw. 1991. Third World urbanization: Dimensions, theories, and determinants. *Annual Review of Sociology* 17:467-501.

Keely, Charles B., and Bao Nga Tran. 1989. Remittances from labor migration: Evaluations, performance and implications. *International Migration Review* 23:500-525.

Kelley, Allen C., and Jeffrey R. Williamson. 1984. *What drives Third World city growth? A dynamic general equilibrium approach.* Princeton, NJ: Princeton University Press.

Kentor, Jeffrey. 1981. Structural determinants of peripheral urbanization: The effects of international dependence. *American Sociological Review* 46:201-211.

Khalifa, Ahmad M., and Mohamed M. Mohieddin. 1988. Cairo. In *The metropolitan era. Vol. 2, Mega-cities,* ed. Mattei Dogan and John D. Kasarda, 235-267. Newbury Park, CA: Sage.

Khan, Mohsin S. 1987. Macroeconomic adjustment in developing countries: A policy perspective. *World Bank Research Observer* 2:23-42.

Kim, Son-Ung, and Peter J. Donaldson. 1979. Dealing with Seoul's population growth: Government plans and their implementation. *Asian Survey* 19:660-673.

Kindleberger, Charles P. 1967. *Europe's postwar growth: The role of labor supply.* Cambridge, MA: Harvard University Press.

———. 1974. The formation of financial centers: A study in comparative economic history. Princeton Studies in International Finance no. 36. Princeton, NJ: International Finance Section, Princeton University.

Kingsley, Thomas. 1988. *Managing urban development: An international perspective.* Washington, DC: Urban Institute.

Kirkby, Richard J. R. 1985. *Urbanisation in China: Town and country in a developing economy 1949-2000 A.D.* London: Croom Helm.

Kobrin, Stephen J. 1977. *Foreign direct investment, industrialization, and social change.* Contemporary Studies in Economic and Financial Analysis no. 9. Greenwich, CT: JAI.

Korcelli, Piotr. 1984. The turnaround of urbanization in developed countries. In *Population distribution, migration and development. International Conference on Population, 1984. Proceedings of the Expert Group on Population Distribution, Migration and Development, Hammamet (Tunisia), 21-25 March 1983,* 349-372. ST/ESA/SER.A/89. New York: Department of International Economic and Social Affairs, United Nations.

Kowarick, Lúcio, and Milton Campanario. 1986. São Paulo: The price of world city status. *Development and Change* 17:159-174.

Kungskulniti, Niapun, Chompusakdi Pulket, F. DeWolfe Miller, and Kirk R. Smith. 1991. Solid waste scavenger community: An investigation in Bangkok, Thailand. *Asia-Pacific Journal of Public Health* 5(1): 54-65.

Kuznets, Simon Smith. 1966. *Modern economic growth: Rate, structure, and spread.* New Haven, CT: Yale University Press.

Laite, Julian. 1985. Circulatory migration and social differentiation in the Andes. *Labour circulation and the labour process,* ed. Guy Standing, 89-119. An ILO-WEP Study. London: Croom Helm.

Lal, Deepak. 1987. The political economy of liberalization. *World Bank Economic Review* 1:273-299.

Laquian, Aprodicio A. 1981. Review and evaluation of urban accommodationist policies in population redistribution. In *Population distribution policies in development planning; Papers of the United Nations/UNFPA Workshop on Population Distribution Policies in Development Planning, Bangkok, 4-13 September 1979,* 101-112. Population Studies no. 75. ST/ESA/SER.A/75. New York: Department of International Economic and Social Affairs, United Nations.

Laszlo, Ervin. 1986. Transformation dynamics in metropolitan systems. In *The metropolis in transition,* ed. Ervin Galantay, 113-121. An ICUS Book. New York: Paragon.

Lee, Barbara W., and John Nellis. 1990. Enterprise reform and privatization in socialist economies. World Bank Discussion Papers no. 104. Washington, DC: World Bank.

Lee, Everett S. 1966. A theory of migration. *Demography* 3:47-57.

Lee, K. S. 1985. *An evaluation of decentralization policies in light of changing location patterns of employment in the Seoul region.* Water Supply and Urban Development Discussion Paper no. 60. Washington, DC: World Bank.

Lesthaeghe, Ron, Georgia Kaufmann, and Dominique Meekers. 1989. The nuptiality regimes in Sub-Saharan Africa. In *Reproduction and social organization in Sub-Saharan Africa,* ed. Ron Lesthaeghe, 238-337. Berkeley: University of California Press.

Levine, David. 1977. *Family formation in an age of nascent capitalism.* New York: Academic Press.

Levy, Hernán, and Aurelio Menéndez. 1990. *Privatization in transport: Contracting out the provision of passenger railway services in Thailand.* EDI Working Papers. Economic Development Institute of the World Bank. Washington, DC: Infrastructure and Urban Development Division, World Bank.

Levy, Marion J., Jr. 1968. *The family revolution in modern China.* New York: Atheneum.

Lewis, W. Arthur. 1954. Development with unlimited supplies of labour. *Manchester School of Economics and Social Studies* 22:139-191.

Liang, Zhixin. 1989. Water shortages in Chinese cities. In *International Symposium 28 Aug-5 Sep 1989. Integrated Water Management and Conservation in Urban Areas. Proceedings of Conference,* 56-68. Nagoya, Japan: International Hydrological Programme.

Lim, E. R., and John Schilling. 1980. Thailand: Toward a development strategy of full participation. World Bank Country Study. Washington, DC: World Bank.

Lim, Linda Y. C. 1983. Singapore's success: The myth of the free market economy. *Asian Survey* 23:752-764.

Lindh, Gunnar. 1983. *Water and the city.* Paris: UNESCO.

Lindh, Gunnar, and J. Niemczynowicz. 1989. Urban water problems in the humid tropics. Paper read at the International Colloquium on the Development of Hydrologic and Water Management Strategies in the Humid Tropics, 15-22 July, Townsville, Australia.

Lipton, Michael. 1982. Migration from rural areas of poor countries: The impact on rural productivity and income distribution. *Migration and the labor market in developing countries,* ed. Richard H. Sabot, 191-228. Boulder, CO: Westview.

Little, I. M. D. 1987. Small manufacturing enterprises in developing countries. *World Bank Economic Review* 1:203-235.

Lo, F. C., K. Salih, and M. Douglass. 1978. Uneven development, rural-urban transformation, and regional development alternatives in Asia. Report submitted to the Seminar on Rural-Urban Transformation and Regional Development Planning, Nov., Japan.

Logan, M. I., and K. Salih. 1982. Implications of international economic adjustments for national urban development and policies. Paper read at the Conference on Urbanization and National Development, East-West Center, Jan., Honolulu.

London, Bruce, and David A. Smith. 1987. Urban bias, dependence, and economic stagnation in non-core nations. Paper presented at the annual meeting of the American Sociological Association, 17-21 Aug., Chicago.

Long, Norman, and Bryan R. Roberts. 1984. *Miners, peasants, and entrepreneurs: Regional development in the central highlands of Peru.* Cambridge: Cambridge University Press.

Low, Kwai Sim. 1989. Urbanization and urban water problems: A case of unsustainable development in ASEAN countries. Paper read at the International Colloquium on the Development of Hydrologic and Water Management Strategies in the Humid Tropics, 15-22 July, Townsville, Australia.

Low, Linda. 1988. Privatization in Singapore. In *Privatization in less developed countries,* ed. Paul Cook and Colin Kirkpatrick, 259-280. New York: St. Martin's.

Lubell, Harold. 1984. Third World urbanization and international assistance. *Urban Studies* 21:1-13.

Maaruf, Annuar bin. 1987. Malaysia country paper. In *Urban Policy Issues,* 529-574. Manila: Asian Development Bank.

Mabogunje, Akin L. 1978. On developing and development 1977. *University of Ibadan Lectures,* 65-99. Ibadan, Nigeria.

McCullough, James S. 1984. Institutional development for local authorities—Financial management consultant's report. Colombo, Sri Lanka: Ministry of Local Government, Housing and Construction.

McCullough, James S., and Thomas H. Steubner. 1985. Project inception report: Management support for Town Panchayats Project. Kathmandu, Nepal: Ministry of Panchayat and Local Development.

McGee, Terence G. 1979. The poverty syndrome: Making out in the Southeast Asian city. In *Casual work and poverty in Third World cities,* ed. Ray J. Bromley and Chris Gerry, 45-68. Chichester, UK: John Wiley.

McGee, Terence G., and Y. M. Yeung. 1977. *Hawkers in Southeast Asian cities: Planning for the bazaar economy.* Ottawa: International Development Research Centre.

Mackenzie, Colin. 1987. The supply lines to Shanghai. *CERES—The FAO Review* 20(5): 16-20.

McNulty, Michael L., and Isaac Ayinde Adalemo. 1988. Lagos. In *The metropolis era.* Vol. 2, *Mega-Cities,* ed. Mattei Dogan and John D. Kasarda, 212-234. Newbury Park, CA: Sage.

Mahbob, Sulaiman bin. 1986. Urbanization of the Malays in Peninsular Malaysia, 1970-1980. Ph.D. diss., Syracuse University.

Mantra, Ida Bagus. 1981. *Population movement in wet rice communities: A case study of two dukuh in Yogyakarta Special Region.* Yogyakarta, Indonesia: Gadjah Mada University Press.

Marmora, Lelio, and Jorge Gurrieri. 1988. *Return to Rio de la Plata: Response to the return of exiles to Argentina and Uruguay.* Washington, DC: Hemispheric Migration Project, Center for Immigration Policy and Refugee Assistance, Georgetown University.

Marshall, Ray. 1988. Jobs: The shifting structure of global employment. *Growth, exports and jobs in a changing world economy—Agenda 1988,* ed. John W. Sewell and Stuart K. Tucker, 167-194. U.S.-Third World Policy Perspectives no. 9. New Brunswick, NJ: Transaction Books.

Martin, Linda, G. 1989. Determinants of living arrangements of the elderly in Fiji, Korea, Malaysia, and the Philippines. *Demography* 26:627-643.

———. 1990. Changing intergenerational family relations in East Asia. In *World population: Approaching the year 2000,* ed. Samuel H. Preston. *Annals of the American Academy of Political and Social Science* 510:102-114.

Martin, Linda G., and Noriko O. Tsuya. 1989. Interactions of middle-aged Japanese with their parents. Paper read at the annual meeting of the Population Association of America, 30 Mar.-1 Apr., Baltimore, MD.

Mason, Karen O., and Anju Malhotra Taj. 1987. Differences between women's and men's reproductive goals in developing countries. *Population and Development Review* 13:611-638.

Massey, Douglas S. 1988. Economic development and international migration in comparative perspective. *Population and Development Review* 14:383-413.

Massey, Douglas S., Rafael Alarcón, Jorge Durand, and Humberto González. 1987. *Return to Aztlan: The social process of international migration from Western Mexico.* Berkeley: University of California Press.

Mathur, Om Prakash. 1989. Urban services and the private sector: Realities and issues. New Delhi: National Institute of Urban Affairs. Typescript.

Mattos, Carlos A. de. 1982. The limits of the possible in regional planning. *CEPAL Review* [Santiago] 18:69-92.

Mayo, S., S. Malpezzi, and D. J. Gross. 1986. Shelter strategies for the urban poor in developing countries. Paper read at the annual meeting of the American Real Estate and Urban Economics Association, New York.

Mehta, R. S. 1982. Problems of shelter, water supply and sanitation in large urban areas. In *Environment and development in Asia and the Pacific: Experience and prospects,* 227-258. UNEP Reports and Proceedings Series no. 6. Nairobi: UNEP [United Nations Environmental Programme].

Meléndez, Gustavo Ponce. 1987. Markets for the Mexican megalopolis. *CERES—The FAO Review* 20(5): 21-26.

Mera, Koichi. 1973. On the urban agglomeration and economic efficiency. *Economic Development and Cultural Change* 21:309-324.

———. 1981. Population distribution policies: The need for caution. In *Population distribution policies in development planning; Papers of the United Nations/UNFOA Workshop on Population Distribution Policies in Development Planning, Bangkok, 4-13 September 1979,* 33-41. Population Studies no. 75. ST/ESA/SER.A/75. New York: Department of International Economic and Social Affairs, United Nations.

Mera, Koichi, and H. Shishido. 1983. Cross-sectional analysis of urbanization and socioeconomic development in the developing world. Washington, DC: World Bank.

Meyer, David R. 1986a. System of cities dynamics in newly industrializing nations. *Studies in Comparative International Development* 21(1): 3-22.

———. 1986b. The world system of cities: Relations between international financial metropolises and South American cities. *Social Forces* 64:553-581.

Mills, Edwin S., and Charles M. Becker 1986. *Studies in Indian urban development.* A World Bank Research Publication. Washington, DC: Oxford University Press for the World Bank.

Mills, Edwin S., and H. S. Kim. 1986. *An economic analysis of Seoul's greenbelt.* Seoul: Korea Research Institute of Human Settlements.

Minis, Henry P., Jr., and Sally S. Johnson. 1982. *Case study of financial management practices in Tunisia.* Working Paper. Research Triangle Park, NC: Research Triangle Institute.

Miró, Carmen A., and Joseph E. Potter. 1980. *Population policy: Research priorities in the developing world.* Report of the International Review Group of Social Science Research on Population and Development. New York: St. Martin's.

Modell, John. 1978. Patterns of consumption, acculturation, and family income strategies in late nineteenth-century America. In *Family and population in nineteenth-century America,* ed. Tamara K. Hareven and Maris A. Vinovskis, 206-240. Princeton, NJ: Princeton University Press.

Modell, John, and Tamara K. Hareven. 1973. Urbanization and the malleable household: An examination of boarding and lodging in American families. *Journal of Marriage and the Family* 35:467-479.

Mohan, Rakesh. 1979. Population, income, and employment in a developing metropolis: A spatial analysis of Bogotá, Colombia. Washington, DC: City Study Research Project, World Bank.

Moir, Hazel. 1977. Dynamic relationships between labor force structure, urbanization, and development. *Economic Development and Cultural Change* 26:25-41.

Montagu-Pollack, Matthew. 1990. Privatization: What went wrong. *Asian Business* 26(8): 32-39.

Montgomery, Mark R., and Ed Brown. 1986. Migration and urbanization in Sub-Saharan Africa. Population, Health and Nutrition Department Paper no. 86-21. Washington, DC: World Bank.

Moran, Theodore H. 1985. Multinational corporations and the developing countries: An analytical overview. In *Multinational corporations: The political economy of foreign direct investment,* ed. Theodore H. Moran, 1-23. Lexington, MA: Lexington Books.

Morgan, S. Philip, and Kiyosi Hirosima. 1983. The persistence of extended family residence in Japan: Anachronism or alternative strategy? *American Sociological Review* 48:269-281.

Morgan, S. Philip, Ronald R. Rindfuss, and Allan M. Parnell. 1984. Modern fertility patterns: Contrasts between the United States and Japan. *Population and Development Review* 10:19-40.

Mukherji, Shekhar. 1985. The process of wage labour circulation in Northern India. *Labour circulation and the labour process,* ed. Guy Standing, 252-289. An ILO-WEP Study. London: Croom Helm.

Murillo, Gabriel, and Gabriel Silva. 1984. La migración de los trabajadores colombianos a Venezuela: Antecedentes y perspectivas. In *Memorias del Congreso Latinoamericano de Población y Desarrollo. Celebrado en la Ciudad de México del 8 al 10 de noviembre de 1983,* vol. 2, 809-830. Mexico City: El Colegio de México, UNAM [Universidad Nacional Autónoma de México], PISPAL [Programa de Investigaciones Sociales sobre Población en América Latina].

Murray, Colin. 1981. *Families divided: The impact of migration in Lesotho.* African Studies Series no. 29. Cambridge, UK: Cambridge University Press.

Musgrove, Philip. 1980. Household size and composition, employment, and poverty in urban Latin America. *Economic Development and Cultural Change* 28:249-266.

National Academy of Public Administration. [Lester M. Salamon, Alan K. Campbell, Lawrence J. Korb, John J. Lordan, Gerald H. Miller, Ronald C. Moe, Brian O'Connell, Harold Seidman, and Dwight Waldo, panel members]. 1989. *Privatization: The Challenge to Public Management.* Washington, DC: National Academy of Public Administration.

National Institute of Urban Affairs. 1987. *Structure and performance of informal enterprises: A study of four cities.* Research Study Series no. 19. New Delhi.

Neck, Philip A., ed. 1977. *Small enterprise development: Policies and programmes.* Management Development Series no. 14. Geneva: International Labour Office.

Nellis, John R., and Sunita Kikeri. 1989. Public enterprise reform: Privatization and the World Bank. *World Development* 17:659-672.

Nelson, Joan M. 1976. Sojourners versus new urbanites: Causes and consequences of temporary versus permanent cityward migration in developing countries. *Economic Development and Cultural Change* 24:721-757.

Noer, Asril. 1985. The agglomeration of manufacturing industries in Indonesia's largest cities, with special focus on domestic and foreign manufacturing investments. Ph.D. diss., Syracuse University.

Oberai, A. S. 1978. *Changes in the structure of employment with economic development.* Geneva: International Labour Office.

———. 1987. *Migration, urbanisation and development.* Background Papers for Training in Population no. 5, Human Resources and Development Planning. Geneva: World Employment Programme, International Labour Organisation.

———. 1989. Rapid population growth, employment and housing in mega-cities in developing countries. In *International Population Conference/Congres international de la population, New Delhi, September/Septembre 20-27, 1987,* vol. 2, 187-199. Liège, Belgium: International Union for the Scientific Study of Population.

O'Connor, Anthony. 1986. The emerging metropolis in tropical Africa. *The metropolis in transition,* ed. Ervin Y. Galantay, 17-32. An ICUS Book. New York: Paragon.

OECD. *See* Organisation for Economic Cooperation and Development.

Oman, Charles. 1984. *New forms of international investment in developing countries.* Paris: Organisation for Economic Cooperation and Development.

Omran, Abdel R. 1971. The epidemiologic transition: A theory of the epidemiology of population change. *Milbank Memorial Fund Quarterly* 49(4, pt. 1): 509-538.

Onibokun, Adepoju G. 1989. Urban growth and urban management in Nigeria. In *African cities in crisis: Managing rapid urban growth,* ed. Richard E. Stren and Rodney R. White, 69-111. Boulder, CO: Westview.

Organisation for Economic Cooperation and Development. [1973]. *Flow of resources to developing countries.* [Paris].

———. 1979. *Migration, growth and development.* Paris.

Pack, Janet Rothenberg. 1989. Privatization and cost reduction. *Policy Sciences* 22:1-25.

PADCO [Planning and Development Cooperative]. 1982. *National Urban Policy Study—Egypt.* 2 vols. Washington, DC.

———. 1991. *India: Public-Private Partnerships in Land Development.* New Delhi: U.S. Agency for International Development.

PADCO-DACREA. 1985. *Analysis of urban services, standards, technologies and costs.* Jakarta: National Urban Development Strategy Project.

Padmopranoto, Sugiarso. 1987. Indonesia country paper. In *Urban Policy Issues,* 431-476. Manila: Asian Development Bank.

Page, Hilary J. 1989. Childrearing versus childbearing: Coresidence of mother and child in Sub-Saharan Africa. In *Reproduction and social organization in Sub-Saharan Africa,* ed. Ron J. Lesthaeghe, 401-441. Berkeley: University of California Press.

Pakkasem, Phisit. 1987. Thailand country paper. In *Urban Policy Issues,* 775-812. Manila: Asian Development Bank.

Palmore, James A., Robert E. Klein, and Ariffin bin Marzuki. 1970. Class and family in a modernizing society. *American Journal of Sociology* 76:375-398.

Papanek, Gustav F. 1973. Aid, foreign private investment, savings, and growth in less developed countries. *Journal of Political Economy* 81:120-130.

Park, Se-Il. 1988. Labor issues in Korea's future. *World Development* 16:99-119.

Pastor, Robert A. 1985. Introduction: The policy challenge. In *Migration and development in the Caribbean: The unexplored connection,* ed. Robert A. Pastor, 1-39. Boulder, CO: Westview.

Peattie, Lisa R. 1975. "Tertiarization" and urban poverty in Latin America. In *Latin American Urban Research.* Vol. 5, *Urbanization and inequality. The political economy of urban and rural development in Latin America,* ed. Wayne A. Cornelius and Felicity M. Trueblood, 109-123. Beverly Hills, CA: Sage.

———. 1987. An idea in good currency and how it grew: The informal sector. *World Development* 15:851-860.

Perlman, Janice E., and Bruce Schearer. 1986. Migration and population distribution trends and policies and the urban future. Paper presented at the International Conference on Population and the Urban Future, Barcelona.

Pernia, Ernesto del Mar. 1988. Urbanization and spatial development in the Asian and Pacific Region: Trends and issues. *Asian Development Review* 6:86-105.

Pessar, Patricia R. 1988. Introduction: Migration myths and new realities. In *When borders don't divide: Labor migration and refugee movements in the Americas,* ed. Patricia R. Pessar, 1-7. Staten Island, NY: Center for Migration Studies.

Peterson, Linda S., and Robert Warren. 1989. Determinants of unauthorized migration to the United States. Paper read at the annual meeting of the Population Association of America, 30 Mar.- 1 Apr., Baltimore, MD.

Peto, Richard. 1980. Distorting the epidemiology of cancer: The need for a more balanced overview. *Nature* 284:297-300.

Phantumvanit, D., and W. Liengcharernsit. 1989. Coming to terms with Bangkok's environmental problems. *Environment and Urbanization* 1:31-39.

Piché, Victor L., Joel N. Gregory, and Sidiki P. Coulibaly. 1980. Vers une explication des courants migratoires voltaiques. *Labour, Capital and Society* 13:77-103.

Piore, Michael J. 1979. *Birds of passage: Migrant labor and industrial societies.* Cambridge, UK: Cambridge University Press.

Popenoe, David. 1988. *Disturbing the nest: Family change and decline in modern societies.* New York: Aldine de Gruyter.

Portes, Alejandro. 1979. Illegal immigration and the international system: Lessons from recent legal Mexican immigrants to the United States. *Social Problems* 26:425-438.

Portes, Alejandro, and Robert L. Bach. 1985. *Latin journey: Cuban and Mexican immigrants in the United States.* Berkeley: University of California Press.

Portes, Alejandro, and József Böröcz. 1989. Contemporary immigration: Theoretical perspectives on its determinants and modes of incorporation. *International Migration Review* 23:606-630.

Portes, Alejandro, Manuel Castells, and Lauren A. Benton. 1989. *The informal economy: Studies in advanced and less developed countries.* Baltimore, MD: Johns Hopkins University Press.

Portes, Alejandro, and Michael Johns. 1986. Class structure and spatial polarization: An assessment of recent urban trends in Latin America. *Tijdschrift voor Economische en Sociale Geografie* 77:378-388.

Portes, Alejandro, and John Walton. 1981. *Labor, class and the international system.* Studies in Social Discontinuity. New York: Academic Press.

Preston, Samuel H. 1976. *Mortality patterns in national populations: With special reference to recorded causes of death.* New York: Academic Press.

———. 1985. Mortality and development revisited. In *Quantitative studies of mortality decline in the developing world,* ed. Julie DaVanzo, Jean-Pierre Habicht, Ken Hill, and Samuel H. Preston, 97-122. World Bank Staff Working Papers no. 683; Population and Development Series no. 8. Washington, DC: World Bank.

Preston, Samuel H., and Michael A. Strong. 1986. Effects of mortality declines on marriage patterns in developing countries. In *Consequences of mortality trends and differentials in developing countries,* 88-100. Population Studies no. 95. ST/ESA/SER.A/95. New York: Department of International Economic and Social Affairs, United Nations.

Prost, A. 1989. The management of water resources, development and human health. Paper read at the International Colloquium on the Development of Hydrologic and Water Management Strategies in the Humid Tropics, 15-22 July, Townsville, Australia.

Prothero, R. Mansell. 1987. Populations on the move. *Third World Quarterly* 9:1282-1310.

Prothero, R. Mansell, and Murray Chapman. 1985. *Circulation in Third World countries.* London: Routledge & Kegan Paul.

Qutub, S. A., and Harry W. Richardson. 1986. The costs of urbanization: A case study of Pakistan. *Environment and Planning A* 18:1089-1113.

Radloff, Scott R. 1982. An evaluation of traditional approaches to measuring migration based on the Malaysian Family Life Survey. Rand Corporation Working Paper. Santa Monica, CA.

Ralston, Lenore, James Anderson, and Elizabeth Colson. 1981. Voluntary efforts in decentralized management: Opportunities and constraints in rural development. Research Series no. 53. Institute of International Studies, University of California, Berkeley.

Ramos, Josephina. 1987. Philippines country paper. In *Urban Policy Issues,* 687-749. Manila: Asian Development Bank.

Ravenstein, Ernest George. 1885. On the laws of migration. *Journal of the Statistical Society* 48(pt. 2): 167-235.

Reichert, Joshua S. 1981. The migrant syndrome: Seasonal U.S. wage labor and rural development in central Mexico. *Human Organization* 40:56-66.

Rempel, Henry. 1981. Rural-urban labor migration and urban unemployment in Kenya. RR-81-24. Laxenburg, Austria: International Institute of Applied Systems Analysis.

Renaud, Bertrand. 1981. *National urbanization policy in developing countries.* New York: Oxford University Press for the World Bank.

———. 1987. Another look at housing finance in developing countries. *Cities* 4:28-34.

Research and Documentation Center for Manpower and Development. 1988. Labor in the public service: Indonesian report. Jakarta. Typescript.

Richardson, Harry W. 1973. *Economics of urban size.* [Farnborough, UK]: Saxon House; [Lexington, MA]: Lexington.

———. 1980. Bombay city study. Washington, DC: World Bank.

———. 1981. Defining urban population distribution goals in development planning. In *Population distribution policies in development planning; Papers of the United Nations/UNFPA Workshop on Population Distribution Policies in Development Planning, Bangkok, 4-13 September 1979,* 7-18. Population Studies no. 75. ST/ESA/SER.A/75. New York: Department of International Economic and Social Affairs, United Nations.

———. 1984a. *The macroeconomic rationale for a medium-size cities project.* Washington, DC: World Bank.

———. 1984b. Planning strategies and policies for Metropolitan Lima. *Third World Planning Review* 6:123-137.

———. 1984c. Population distribution policies. In *Population distribution, migration and development. Proceedings of the Expert Group on Population Distribution, Migration and Development, Hammamet (Tunisia), 21-25 March 1983,* 262-293. ST/ESA/SER.A/89. New York: Department of International Economic and Social Affairs, United Nations.

———. [1984d]. Spatial strategies and infrastructure planning in the metropolitan areas of Bombay and Calcutta. In *Spatial, environmental and resource policy in the developing countries,* ed. Manas Chatterjee, Peter Nijkamp, T. R. Lakshmann, and C. R. Pathak, 113-139. Aldershot, UK: Gower.

———. 1985. *Population distribution and urbanization: A review of policy options.* HS/73/85. Nairobi: United Nations Centre for Human Settlements.

———. 1987a. The costs of urbanization: A four-country comparison. *Economic Development and Cultural Change* 35:561-580.

———. 1987b. Whither national urban policy in developing countries? *Urban Studies* 24:227-244.

———. 1987c. Spatial strategies, the settlement pattern, and shelter and services policies. In *Shelter, settlement and development,* ed. Lloyd Rodwin, 207-235. Boston: Allen & Unwin.

———. 1989a. The big, bad city: Mega-city myth? *Third World Planning Review* 11:355-372.

———. 1989b. National urban policies and the costs and benefits of urbanization. In *World population trends and their impact on economic development,* ed. Dominick Salvatore, 95-106. Contributions in Economics and Economic History no. 82. New York: Greenwood.

Richardson, Harry W., and Myong-Chan Hwang. 1987. *Urban and regional policies in Korea and international experiences.* Seoul: Kon-Kuk University Press in association with Korea Research Institute for Human Settlements.

Roberts, Bryan R. 1976. The provincial urban system and the process of dependency. In *Current perspectives in Latin American urban research,* ed. Alejandro Portes and Harley L. Browning, 99-130. Austin: Institute of Latin American Studies, University of Texas.

———. 1989. Employment structure, life cycle, and life chances: Formal and informal sectors in Guadalajara. In *The informal economy: Studies in advanced and less developed countries,* ed. Alejandro Portes, Manuel Castells, and Lauren Benton, 41-59. Baltimore, MD: Johns Hopkins University Press.

Robinson, Warren C. 1987. Implicit policies and the urban bias as factors affecting urbanization. In *Urbanization and urban policies in Pacific Asia,* ed. Roland J. Fuchs, Gavin W. Jones, and Ernesto del Mar Pernia with Sandra E. Ward, 169-182. Westview Special Studies on East Asia. Boulder, CO: Westview in association with the East-West Population Institute, East-West Center, Honolulu, and National Centre for Development Studies, Australian National University, Canberra.

Rocca, Carlos A. 1970. Appendix 2. Productivity in Brazilian manufacturing. In *Brazil: Industrialization and trade policies,* ed. Joel Bergsman, 222-241. Industry and Trade in Some Developing Countries. London: Oxford University Press on behalf of the Development Centre of the Organisation for Economic Cooperation and Development.

Rogerson, C. M. 1982. Multinational corporations in Southern Africa: A spatial perspective. *The geography of multinationals: Studies in the spatial development and economic consequences of multinational corporations,* ed. Michael J. Taylor and Nigel J. Thrift, 179-220. New York: St. Martin's.

Rondinelli, Dennis A. 1983. *Secondary cities in developing countries: Policies for diffusing urbanization.* Sage Library of Social Research no. 145. Beverly Hills, CA: Sage.

———. 1984. Land-development policy in South Korea. *Geographical Review* 74:425-440.

———. 1986. Extending urban services in developing countries: Policy options and organizational choices. *Public Administration and Development* 6(1): 1-21.

———. 1987. Cities as agricultural markets. *Geographical Review* 77:408-420.

———. 1988. Giant and secondary city growth in Africa. In *The metropolis era.* Vol. 1, *A world of giant cities,* ed. Mattei Dogan and John D. Kasarda, 291-321. Newbury Park, CA: Sage.

———. 1989. Balanced patterns of urbanization in developing countries: The concept and reality. Paper read at the International Workshop on Urbanization and Population Distribution Policies in Asia, Honolulu.

———. 1990a. Housing the urban poor in developing countries: Magnitude of housing deficiencies and failure of conventional strategies are world-wide problems. *American Journal of Economics and Sociology* 49:153-166.

———. 1990b. Policies for balanced urban development in Asia: Concepts and reality. *Regional Development Dialogue* [Nagoya, Japan] 11:23-51.

———. 1991. Asian urban development policies in the 1990s: From growth control to urban diffusion. *World Development* 19:791-803.

———. 1992. *Development projects as policy experiments: An adaptive approach to development administration.* 2d rev. ed. London: Routledge & Kegan Paul.

Rondinelli, Dennis A., and John D. Kasarda. 1991. Privatizing public services in developing countries: What do we know? *Business in the Contemporary World* 3(2): 102-113.

Rondinelli, Dennis A., John Middleton, and Adriaan M. Verspoor. 1990. *Planning Education Reforms in Developing Countries.* Durham, NC: Duke University Press.

Rondinelli, Dennis A., John R. Nellis, and G. Shabbir Cheema. 1983. *Decentralization in developing countries: A review of recent experience.* World Bank Staff Working Papers no. 581. Washington, DC: World Bank.

Rondinelli, Dennis A., and Kenneth Ruddle. 1978. *Urbanization and rural development: A spatial policy for equitable growth.* New York: Praeger.

Roth, Gabriel J. 1987. *The private provision of public services in developing countries.* New York: Oxford University Press.

Ruggles, Steven. 1987. *Prolonged connections: The rise of the extended family in nineteenth-century England and America.* Madison: University of Wisconsin Press.

Ryder, Norman B. 1980. Where do babies come from? In *Sociological theory and research: A critical appraisal,* ed. Hubert M. Blalock, Jr., 189-202. New York: Free Press.

Sachs, Ignacy. 1988. Vulnerability of giant cities and the life lottery. In *The metropolis era.* Vol. 1, *A world of giant cities,* ed. Mattei Dogan and John D. Kasarda, 337-350. Newbury Park, CA: Sage.

Saith, Ashwani. 1989. Macro-economic issues in international labour migration: A review. In *To the Gulf and back: Studies on the economic impact of Asian labour migration,*

ed. Rashid Amjad, 28-54. RAS/85/009. Geneva: Asian Employment Programme, International Labour Organisation.
Salaff, Janet W. 1976. The status of unmarried Hong Kong women and the social factors contributing to their delayed marriage. *Population Studies* 30:391-412.
Salt, John. 1989. A comparative overview of international trends and types, 1950-80, *International Migration Review* 23:431-456.
Sanderson, Susan Walsh, Gregory Williams, Timothy Ballenger, and Brian J. L. Berry. 1987. Impacts of computer-aided manufacturing on offshore assembly and future manufacturing locations. *Regional Studies* 21:131-142.
Santiago, Asteya M. 1989. Subdivision-based approach to central business and commercial district development—The case of Makati, Metro Manila. *Regional Development Dialogue* 10(4): 20-37.
Savas, Emanuel S. 1982. *Privatizing the public sector.* Chatham, NJ: Chatham.
Schteingart, Martha. 1989. The environmental problems associated with urban development in Mexico City. *Environment and Urbanization* 1:40-50.
Seccombe, Ian J., and Allan M. Findlay. 1989. The consequences of temporary emigration and remittance expenditure from rural and urban settlements: Evidence from Jordan. In *The impact of international migration on developing countries,* ed. Reginald T. Appleyard, 109-125. Paris: Development Centre, Organisation for Economic Cooperation and Development.
Sethuraman, S. V. 1976. The urban informal sector: Concept, measurement and policy. *International Labour Review* 114:69-81.
———. 1985. The informal sector in Indonesia: Policies and prospects. *International Labour Review* 124:719-735.
Shanghai Population Information Centre. 1989. Floating population: 50 million. *China Population Research Leads* no. 5:4.
Shantakumar, G. 1984. Human resources development in Singapore. In *Singapore: Twenty five years of development,* ed. You Poh Seng and Lim Chong Yah, 165-188. [Singapore]: Nan Yang Xing Zhou Lianhe Zaobao.
Sharma, M. L. 1986. *Role of groundwater in urban water supplies of Bangkok, Thailand, and Jakarta, Indonesia.* Working Paper, Environment and Policy Institute, East-West Center, Honolulu.
Shaw, R. Paul. 1975. *Migration theory and fact: A review and bibliography of current literature.* Philadelphia: Regional Science Institute.
Shirley, Mary M. 1983. *Managing state-owned enterprises.* World Bank Staff Working Papers no. 577. Management and Development Series no. 4. Washington, DC: World Bank.
Shukla, Vibhooti. 1984. The productivity of Indian cities and some implications for development policy. Ph.D. diss., Princeton University.
Simmons, Alan B. 1979. Slowing metropolitan city growth in Asia: Policies, programs, and results. *Population and Development Review* 5:87-104.
———. 1981. A review and evaluation of attempts to constrain migration to selected urban centres and regions. In *Population distribution policies in development planning; Papers of the United Nations/UNFPA Workshop on Population Distribution Policies in Development Planning, Bangkok, 4-13 September 1979,* 87-100. Population Studies no. 75. ST/ESA/SER.A/75. New York: Department of International Economic and Social Affairs, United Nations.

———. 1984. Migration and rural development: Conceptual approaches, research findings and policy issues. In *Population Distribution, migration and development,* 156-192. ST/ESA/Ser.A./89. New York: Department of International Economics and Social Affairs, United Nations.

Singh, Ajit. 1989. *Urbanisation, poverty and employment: The large metropolis in the Third World.* Population and Labour Policies Programme, Working Paper no. 165. World Employment Program. Geneva: International Labour Office.

Singh, R. D. 1988. The multinationals' economic penetration, growth, industrial output and domestic savings in developing countries: Another look. *Journal of Development Studies* 25:55-82.

Singh, Susheela. 1980. Background characteristics used in WFS surveys. *World Fertility Survey Comparative Studies,* no. 4.

Sivaramakrishnan, K. C., and Leslie Green. 1986. *Metropolitan management: The Asian experience.* New York: Oxford University Press for the Economic Development Institute of the World Bank.

Smart, J., and V. Teodosio. 1983. Skills and earnings: Issues in the developmental impact of Middle East employment on the Philippines. Paper read at the Conference on Asian Migration to the Middle East, East-West Population Institute, Honolulu.

Smil, Vaclav. 1984. *The bad earth: Environmental degradation in China.* Armonk, NY: Sharpe.

Smith, Carol A. 1985. Theories and measures of urban primacy: A critique. *Urbanization in the world-economy,* ed. Michael Timberlake, 87-117. Orlando, FL: Academic Press.

Smith, David A. 1985. International dependence and urbanization in East Asia: Implications for planning. *Population Research and Policy Review* 4:203-233.

Smith, David P., and Benoit Ferry. 1984. Correlates of breastfeeding. *World Fertility Survey Comparative Studies,* no. 41.

Smith, Herbert L. 1989. Integrating theory and research on the institutional determinants of fertility. *Demography* 26:171-184.

Smith, Kirk R. 1988a. Air pollution: Assessing total exposure in the United States. *Environment* 30(8): 10-15, 33-38.

———. 1988b. Air pollution: Assessing total exposure in developing countries. *Environment* 30(10): 16-20, 28-35.

———. 1990a. Indoor air quality and the pollution transition. In *Indoor air quality,* ed. Hitoshi Kasuga, 448-456. Berlin: Springer.

———. 1990b. The risk transition. *International Environmental Affairs* 2:227-251.

Smith, Peter C. 1983. The impact of age at marriage and proportions marrying on fertility. In *Determinants of fertility in developing countries.* Vol. 2, *Fertility regulation and institutional influences,* ed. Rodolfo A. Bulatao and Ronald D. Lee with Paula E. Hollerbach and John Bongaarts, 473-531. New York: Academic Press.

Smith, Peter C., Siew-Ean Khoo, and Stella P. Go. 1984. The migration of women to cities: A comparative perspective. In *Women in the cities of Asia: Migration and urban adaptation,* ed. James T. Fawcett, Siew-Ean Khoo, and Peter C. Smith, 15-35. Boulder, CO: Westview.

Squire, Lyn. 1981. *Employment policy in developing countries: A survey of issues and evidence.* New York: Oxford University Press for the World Bank.

Stahl, Charles W. 1984. The economic impact of labour emigration. Paper read at the International Union for the Scientific Study of Population Workshop on the Consequences of International Migration, 16-19 July, Canberra, Australia.

Standing, Guy, ed. 1985 *Labour circulation and the labour process.* London: Croom Helm.

Steel, William F. 1977. *Small-scale employment and production in developing countries: Evidence from Ghana.* New York: Praeger.

Sternstein, Larry. 1984. The growth of the population of the world's pre-eminent "primate city": Bangkok at its bicentenary. *Journal of Southeast Asia Studies* 15 (March): 43-68.

Stretton, Alan. 1985. Circular migration, segmented labour markets and efficiency. *Labour circulation and the labour process,* ed. Guy Standing, 290-312. An ILO-WEP Study. London: Croom Helm.

Stichter, Sharon. 1985. *Migrant laborers.* African Society Today. Cambridge, UK: Press Syndicate of the University of Cambridge.

Stier, K. J. 1990. Thai privatization stirs widespread opposition. *Journal of Commerce and Commercial* (16 Feb.): 5A.

Stöhr, Walter B. 1981. Evaluation of some arguments against government intervention to influence territorial population distribution. In *Population distribution policies in development planning; Papers of the United Nations/UNFPA Workshop on Population Distribution Policies in Development Planning, Bangkok, 4-13 September 1979,* 42-49. Population Studies no. 75. ST/ESA/SER.A/75. New York: Department of International Economic and Social Affairs, United Nations.

Stolnitz, George J. 1984. *Urbanisation and rural-to-urban migration in relation to LDC fertility.* Project Paper no. 10. Bloomington: Fertility Determinants Group, Indiana University

Stren, Richard E. 1989. The administration of urban services. In *African cities in crisis: Managing rapid urban growth,* ed. Richard E. Stren and Rodney R. White, 37-67. Boulder, CO: Westview.

Sukthankar, D. M., and P. S. A. Sundaram. 1987. India country paper. In *Urban policy issues,* 393-429. Manila: Asian Development Bank.

Swamy, Gurushri. 1981. *International migrant workers' remittances: Issues and prospects: A background study for world development report 1981.* World Bank Staff Working Papers no. 481. Washington, DC: World Bank.

Tabbarah, Riad B. 1981. Changing patterns of international migration and their demoeconomic implications. In *International Population Conference, Manila, 1981; Proceedings and selected papers,* vol. 4, 181-185. Liège, Belgium: International Union for the Scientific Study of Population.

Taira, Koji. 1969. Urban poverty, ragpickers, and the "Ants' Villa" in Tokyo. *Economic Development and Cultural Change* 17:155-177.

Tan Chee Hock. 1987. Use of performance analysis to determine privatization of street parking. Paper prepared for Course on Urban Finance and Management in East Asia, 7 Sept.-2 Oct., sponsored by National Institute of Public Administration, Malaysia, Economic Development Institute of the World Bank, United Nations Centre for Human Settlements, and the United Nations Development Programme, Kuala Lumpur.

Taylor, Michael J., and Nigel J. Thrift, eds. 1982. *The geography of multinationals: Studies in the spatial development and economic consequences of multinational corporations.* New York: St. Martin's.

Ternent, J. Anthony. 1976. Urban concentration and dispersal: Urban policies in Latin America. In *Development planning and spatial structure,* ed. Alan Gilbert, 290-312. London: John Wiley.

Teune, Henry. 1988. Growth and pathologies of giant cities. In *The metropolis era.* Vol. 1, *A world of giant cities,* ed. Mattei Dogan and John D. Kasarda, 351-376. Newbury Park, CA: Sage.

Thailand. National Statistical Office. 1976. *The survey of migration in Bangkok metropolis 1976.* Bangkok: Office of the Prime Minister.

Thomas, Robert N., and Kevin F. Byrnes. 1974. Intervening opportunities and the migration field of a secondary urban center: The case of Tunja, Colombia. In *Latin America: Search for geographic explanations,* ed. Robert J. Tata, 83-88. Conference of Latin Americanist Geographers, vol. 5. Chapel Hill: C.L.A.G. Publications, Department of Geography, University of North Carolina.

Thomas-Hope, Elizabeth M. 1985. Return migration and its implications for Caribbean development. In *Migration and development in the Caribbean: The unexplored connection,* ed. Robert A. Pastor, 157-177. Boulder, CO: Westview.

Thornton, Arland, Ming-Cheng Chang, and Te-Hsung Sun. 1984. Social and economic change, intergenerational relationships, and family formation in Taiwan. *Demography* 21:475-499.

Thornton, Arland D., and Thomas E. Fricke. 1987. Social change and the family: Comparative perspectives from the West, China, and South Asia. *Sociological Forum* 2:746-779.

Tienda, Marta. 1979. Economic activity of children in Peru: Labor force behavior in rural and urban contexts. *Rural Sociology* 44:370-391.

Timaeus, Ian, and Wendy Graham. 1989. Labor circulation, marriage, and fertility in Southern Africa. In *Reproduction and social organization in Sub-Saharan Africa,* ed. Ron J. Lesthaeghe, 365-400. Berkeley: University of California Press.

Timberlake, Michael. 1985. The world-system perspective and urbanization. *Urbanization in the world-economy,* ed. Michael Timberlake, 3-39. New York: Academic Press.

Todaro, Michael P. 1986. *International migration, domestic unemployment, and urbanization: A three-sector model.* Center for Policy Studies, Working Papers no. 124. New York: Population Council.

Tokman, Victor E. 1978. An exploration into the nature of informal-formal sector relationships. *World Development* 6:1065-1075.

———. 1988. Urban employment: Research and policy in Latin America. *CEPAL Review* [Santiago] no. 34: 109-126.

Townroe, Peter M. 1983. *Location factors in the decentralization of industry: A survey of Metropolitan São Paulo, Brazil.* World Bank Staff Working Papers no. 517. Washington, DC: World Bank.

Tsuruoka, Doug. 1990. Privatised patronage. *Far Eastern Economic Review* 150 (20 Dec.): 42-44.

Tucker, Stuart K., and Carey Durkin Treado. 1988. Statistical annexes: U.S. Third World interdependence. In *Growth, exports and jobs in a changing world economy—Agenda 1988,* ed. John W. Sewell and Stuart K. Tucker, 195-268. U.S.-Third World Policy Perspectives no. 9. New Brunswick, NJ: Transaction Books.

UN. *See* United Nations.

UNCHS. *See* United Nations Centre for Human Settlements.

UNCRD. *See* United Nations Centre for Regional Development.

UNDIESA. *See* United Nations. Department of International Economic and Social Affairs.

UNEP. *See* United Nations Environment Programme.

UNESCAP. *See* United Nations. Economic and Social Commission for Asia and the Pacific.

UNESCAP/CTC. See United Nations. Economic and Social Commission for Asia and the Pacific/Joint Unit on Transnational Corporations.

Ungar Bleier, E. 1988. Impact of the Venezuelan recession on return migration to Colombia: The case of the principal urban sending areas. In *When borders don't divide: Labor migration and refugee movements in the Americas,* ed. Patricia R. Pessar, 73-95. Staten Island, NY: Center for Migration Studies.

[United Nations]. 1984. The International Conference on Population, 1984. *Population and Development Review* 10:755-782.

United Nations. Department of International Economic and Social Affairs. 1980. *Patterns of urban and rural population growth.* Population Studies no. 68. New York.

———. 1981. *Population distribution policies and development planning; Papers of the United Nations/UNFPA Workshop on Population Distribution Policies in Development Planning, Bangkok, 4-13 September 1979.* Population Studies no. 75. ST/ESA/Ser.A/75. New York.

———. 1984. *International Conference on Population 1984, Tunisia. Population Distribution, Migration and Development. Proceedings of the Expert Group on Population Distribution, Migration, and Development, Hammamet (Tunisia), 21-25 March 1983.* ST/ESA/SER.A/89. New York.

———. 1985. *Migration, population growth and employment in metropolitan areas of selected developing countries.* ST/ESA/SER.R/57. New York.

———. 1986a. *Population growth and policies in mega-cities: Bombay.* Population Policy Paper no. 6. ST/ESA/SER.A/67. New York.

———. 1986b. *Population growth and policies in mega-cities: Calcutta.* Population Policy Paper no. 1. ST/ESA/SER.R/61. New York.

———. 1986c. *Population growth and policies in mega-cities: Delhi.* Population Policy Paper no. 7. ST/ESA/SER.A/68. New York.

———. 1986d. *Population growth and policies in mega-cities: Metro Manila.* Population Policy Paper no. 5. ST/ESA/SER.A/65. New York.

———. 1986e. *Population growth and policies in mega-cities: Seoul.* Population Policy Paper no. 4. ST/ESA/SER.A/64. New York.

———. 1986f. *Review and appraisal of the world population plan of action, 1984 report.* ST/ESA/SER.A/99. New York.

———. 1987a. *Population growth and policies in mega-cities: Bangkok.* Population Policy Paper no. 10. ST/ESA/SER.R/72. New York.

———. 1987b. *Population growth and policies in mega-cities: Dhaka.* Population Policy Paper no. 8. ST/ESA/SER.R/69. New York.

———. 1987c. *Population growth and policies in mega-cities: Madras.* Population Policy Paper no. 12. ST/ESA/SER.R/75. New York.

———. 1987d. *The prospects of world urbanization; revised as of 1984-85.* Population Studies no. 101. ST/ESA/SER.A/101. New York.

———. 1988a. *Population growth and policies in mega-cities: Karachi.* Population Policy Paper no. 13. ST/ESA/SER.A/77. New York.

———. 1988b. *World demographic estimates and projections, 1950-2025.* ST/ESA/SER.R/79. New York.

———. 1988c. *World population trends and policies: 1987 monitoring report.* Population Studies no. 103. ST/ESA/SER.A/103. New York.

———. 1989a. *Population growth and policies in mega-cities: Jakarta.* Population Policy Paper no. 18. ST/ESA/SER.A/86. New York.
———. 1989b. *Prospects of world urbanization, 1988.* Population Studies no. 112. ST/ESA/SER.A/112. New York.
———. 1989c. *World population trends and policies: 1989 monitoring report.* ST/ESA/SER.A/ 107. New York.
———. 1990a. *Population growth and policies in mega-cities: Cairo.* Population Policy Paper no. 33. ST/ESA/SER.A/103. New York.
———. 1990b. *Population growth and policies in mega-cities: Mexico City.* Population Policy Paper no. 34. ST/ESA/SER.A/105. New York.
———. 1991. *World urbanization prospects 1990: Estimates and projections of urban and rural populations and of urban agglomerations.* Population Studies no. 121. ST/ESA/SER.A/121. New York.
United Nations. Economic and Social Commission for Asia and the Pacific. 1988. *Internal migration and structural changes in the labor force.* Asian Population Studies Series no. 90. Bangkok.
United Nations. Economic and Social Commission for Asia and the Pacific/Joint Unit on Transnational Corporations. 1986. *Transnational corporations and external financial flows of developing economies in Asia and the Pacific.* ESCAP/UNCTC Publication Series B, no. 10. Bangkok.
United Nations. Economic and Social Commission for Asia and the Pacific. Population Division. 1984. *Urbanization in Thailand and its implications for the family planning programme.* Population Research Leads no. 17. Bangkok.
———. 1989. Issues in privatization. *Development Planning Newsletter* no. 9: 10-15.
United Nations Centre for Human Settlements. 1982. *Survey of slums and squatter settlements.* Development Studies Series no. 1. Dublin: Tycooly International Publishing for United Nations Centre for Human Settlements (Habitat).
———. 1984a. *Environmental aspects of water management in metropolitan areas of developing countries: Issues and guidelines.* HS/42/84. Nairobi.
———. 1984b. *Upgrading of inner-city slums.* HS/CONF/84-1. Nairobi.
———. 1985. *The role of small and intermediate settlements in national development.* Nairobi.
———. 1986. *Rehabilitation of inner city areas: Feasible strategies.* HS/79/85. Nairobi.
———. 1987. *Global report on human settlements.* New York: Oxford University Press for the United Nations Centre for Human Settlements (Habitat).
———. 1988. Research and development. UNCHS/ECA promote local building materials in Africa. *Habitat News* 10(2): 25-31.
———. 1989a. *Analysis of land markets: Analysis and synthesis report.* Draft synopsis. Urban Management Group.
———. 1989b. *Analysis of land markets: Analysis and synthesis report.* Revised draft. Urban Management Group.
United Nations Centre for Regional Development. 1989. Solid waste management for metropolitan development and management in Asia. Abstract. *Regional Development Dialogue* 10(3): Appendix 2.
United Nations Environment Programme. 1982. *Environment and development in Asia and the Pacific: Experience and prospects.* Nairobi.
———. 1987. *Environmental data report.* Prepared for UNEP by the Monitoring and Assessment Research Centre, London, in cooperation with the World Resources Insti-

tute, Washington, DC; International Institute for Environment and Development, London and Washington, DC; and UK Department of the Environment, London. Oxford, UK: Basil Blackwell.

United States. Agency for International Development. 1991. *Urban economies and national development* by George E. Peterson, G. Thomas Kingsley, and Jeffrey P. Telgarsky. Policy and Research Series of the Office of Housing and Urban Programs, Washington, DC.

Unvala, S. P. 1989. Bombay's water supply situation: Drought and migration wreak havoc on limited resource. *Water and Wastewater International* 4(1): 33-37.

USAID. *See* United States. Agency for International Development.

Van de Walle, Nicolas. 1989. Privatization in developing countries: A review of the issues. *World Development* 17:601-615.

van Ginneken, Wouter. 1988. Employment and labor incomes: A cross country analysis (1971-86). In *Trends in employment and labour incomes: Case studies on developing countries,* ed. Wouter van Ginneken, 1-31. Geneva: International Labour Office.

Van Huyck, Alfred P. 1988. Urbanization in the developing countries: Opportunities for United States development cooperation. Paper prepared for the Office of Housing and Urban Programs, U.S. Agency for International Development. Washington, DC.

Verduzco Igartúa, Gustavo. 1990. La migración urbana a Estados Unidos: Un caso del occidente de México. *Estudios Sociológicos* 8(22): 117-139.

Vernon, Raymond. 1971. *Sovereignty at bay: The multinational spread of U.S. enterprises.* New York: Basic Books.

———. 1988. Introduction: The promise and the challenge. In *The promise of privatization: A challenge for U.S. policy,* ed. Raymond Vernon, 1-22. New York: Council on Foreign Relations.

Vernon-Wortzel, Heidi, and Lawrence H. Wortzel. 1989. Privatization: Not the only answer. *World Development* 17:633-641.

Vialet, Joyce, and Barbara McClure. 1980. *Temporary worker programs, background and issues. A report prepared at the request of Senator Edward M. Kennedy, Chairman, Committee on the Judiciary, U.S. Senate, for the use of the Select Commission on Immigration and Refugee Policy.* Washington, DC: U.S. Government Printing Office.

Vichit-Vadakan, Juree. 1984. Small towns and regional urban centres: Reflections on diverting Bangkok-bound migration. In *Equity with growth? Planning perspectives for small towns in developing countries,* ed. H. Detlef Kammeier and Peter J. Swan, 489-492. Bangkok: Asian Institute of Technology.

Vining, Daniel R., Jr. 1986. Population redistribution towards core areas of less developed countries, 1950-1980. *International Regional Science Review* 10:1-45.

Vuylsteke, Charles. 1988. *Techniques of privatization of state-owned enterprises.* Vol. 1, *Methods and implementation.* World Bank Technical Paper no. 88-90. Washington, DC: World Bank.

Watkins, Susan Cotts. 1984. Spinsters. *Journal of Family History* 9:310-325.

White, Rodney R. 1989. The influence of environmental and economic factors on the urban crisis. In *African cities in crisis: Managing rapid urban growth,* ed. Richard E. Stren and Rodney R. White, xiv-19. Boulder, CO: Westview.

Whitney, Joseph B. R. 1991. The waste economy and its dispersed metropolis in China. In *The extended metropolis: Settlement transition in Asia,* ed. Norton S. Ginsburg, Bruce Koppel, and Terence G. McGee, 177-191. Honolulu: University of Hawaii Press.

WHO. *See* World Health Organization.

Williams, Brian T. 1990. Assessing the health impact of urbanization. *World Health Statistics Quarterly* 43:145-152.

Williamson, Jeffrey G. 1988. Migration and urbanization. In *Handbook of development economics,* ed. Hollis Cheney and T. N. Srinivasan, vol 1. Handbooks in Economics no. 9. New York: North Holland.

Wilson, Francis. 1976. International migration in Southern Africa. *International Migration Review* 10:452-488.

Wirth, Louis. 1938. Urbanism as a way of life. *American Journal of Sociology* 44:1-24.

World Bank. 1983a. *World development report 1983.* New York: Oxford University Press for the World Bank.

———. 1983b. *World tables, from the data files of the World Bank.* Vol. 2, *Social data.* Baltimore, MD: Johns Hopkins University Press for the World Bank.

———. 1984. Seoul: The impact of government policies. *Urban Edge. Issues and Innovations* 8(9): 4.

———. 1986. *Urban transport.* Washington, DC.

———. 1988a. "Enabling" shelter strategy urged by U.N. body. *Urban Edge. Issues and Innovations* 12(8): 1-2, 5.

———. 1988b. *World development report 1988.* New York: Oxford University Press for the World Bank.

———. 1989a. Debate over land registration persists. *Urban Edge. Issues and Innovations* 13(7): 1-4.

———. 1989b. *FY89 sector review of urban development operations: Reaching the poor through urban operations.* Infrastructure and Urban Development Department. Washington, DC.

———. 1989c. *World development report 1989.* New York: Oxford University Press for the World Bank.

———. 1990. Structural adjustment and sustainable growth: The urban agenda for the 1990s. Infrastructure and Urban Development Department, Urban Development Division, Washington, DC. Draft typescript.

———. 1991. *World development report 1991.* New York: Oxford University Press for the World Bank.

World Bank. Country Economics Department. 1991. *The reform of public sector management: Lessons from experience.* Policy and Research Series no. 18. Washington, DC.

World Health Organization. 1986. *World health statistics annual, 1986.* Geneva.

———. 1988a. *Urbanization and its implications for child health: Potential for action.* Geneva: WHO in collaboration with the United Nations Environment Programme.

———. 1988b. *Review of progress of the international drinking water supply and sanitation decade, 1981-1990: Eight years of implementation. Report by the Director-General.* Geneva.

Wurzel, P. 1989. Water supply and health in the humid tropics. Paper read at the International Colloquium on the Development of Hydrologic and Water Management Strategies in the Humid Tropics, 15-22 July, Townsville, Australia.

Yeh, Anthony Gar-On. 1990. Public and private partnership in urban redevelopment in Hong Kong. *Third World Planning Review* 12: 361-383.

Yi, Zeng, and James W. Vaupel. 1989. The impact of urbanization and delayed childbearing on population growth and aging in China. *Population and Development Review* 15:425-446.

Yotopoulos, Pan A. 1989. The (rip) tide of privatization: Lessons from Chile. *World Development* 17:683-702.

Yu, Qingkang, and Gu Wenxuan. 1984. In search of an approach to rural-urban integration. *Third World Planning Review* 6:37-46.

Yuan, Yihua. 1989. A study of the contemporary population migration in the People's Republic of China. Research Discussion Paper no. 59. Edmonton, Canada: Population Research Laboratory, Department of Sociology, University of Alberta.

Zelinsky, Wilbur. 1971. The hypothesis of the mobility transition. *Geographical Review* 61:219-249.

———. 1979. The demographic transition: Changing patterns of migration. In *Population science in the service of mankind, Conference on science in the service of life, Vienna, 1979,* 165-189. Liège, Belgium: International Union for the Scientific Study of Population.

Zheng Guizhen, Guo Shenyang, Zhang Yunfan, and Wang Jufen. 1985a. A study on the floating population of Shanghai City proper. *Population Research* [Beijing] 3 (3): 15-22 [in Chinese].

———. 1985b. A study of temporary population movement in Shanghai. Fudan University, Shanghai. Typescript.

Zhu, Baoshu, and Giuxing Wang. 1985. The transformation of rural population and the control of city population. *Shandong Population* 1:59-64 [in Chinese].

Zlotnik, Hania. 1987. The concept of international migration as reflected in data collection systems. *International Migration Review* 4:925-946.

Zolberg, Aristide R. 1989. The next waves: Migration theory for a changing world. *International Migration Review* 3:403-430.

Zoughlami, Younes, and Diana Allsopp. 1985. The demographic characteristics of household populations. *World Fertility Survey Comparative Studies,* no. 45.

Author Index

Subject Index

About the Authors and Editors

Francisco Alba, an economist and demographer, is Professor and Researcher at the Center for Demographic and Urban Studies, El Colegio de México in Mexico City.

Ellen M. Brennan, Acting Chief, Population Policy Section, United Nations Population Division, is the author of the United Nations' series of case studies entitled *Population Growth and Policies in Mega-Cities.*

Ray Bromley is Professor of Geography, Planning, and Latin American Studies at the State University of New York at Albany, where he also directs the Lewis Mumford Center for Urban and Regional Research.

Sally E. Findley is an Associate Professor with the Center for Population and Family Health, Faculty of Medicine of Columbia University, New York.

Sidney Goldstein is G. H. Crooker University Professor and Professor of Sociology at Brown University, Providence, RI, and served as director of Brown's Population Studies and Training Center from 1965 to 1989. In 1990 he was successively a Senior Fellow at the East-West Population Institute, a Scholar in Residence at the Rockefeller Foundation Study Center in Bellagio, Italy, and Senior Visiting Fellow at the Hebrew University of Jerusalem.

John D. Kasarda is Kenan Professor of Business Administration and Sociology at the University of North Carolina at Chapel Hill, director of its Kenan Institute of Private Enterprise at the Kenan-Flagler Business School, and a Fellow of the Carolina Population Center.

Yok-shiu F. Lee is a Research Associate in the Environment and Policy Institute, East-West Center, Honolulu.

S. Philip Morgan is Professor of Sociology, Population Studies Center, University of Pennsylvania, Philadelphia. In 1990 he was a Visiting Fellow in the Department of Demography, Research School of Social Sciences, the Australian National University.

A. S. Oberai is Senior Economist, Employment Planning and Population Branch, Employment and Development Department, International Labour Office, Geneva.

Allan M. Parnell is Assistant Professor of Sociology at Duke University.

Harry W. Richardson holds a joint appointment as Professor in the School of Urban and Regional Planning and the Department of Economics, University of Southern California, Los Angeles.

Dennis A. Rondinelli is Director, International Private Enterprise Development Research Center, Kenan Institute of Private Enterprise, and Professor of Business Administration, Kenan-Flagler Business School, University of North Carolina at Chapel Hill.

Victor Fung-Shuen Sit is a Reader in the Department of Geography and Geology, University of Hong Kong.

Kirk R. Smith is a Research Associate with the Environment and Policy Institute of the East-West Center, Honolulu.